Karen S. Feldman
Arts of Connection

Paradigms

Literature and the Human Sciences

Edited by
Rüdiger Campe · Paul Fleming

Editorial Board
Eva Geulen · Rüdiger Görner · Barbara Hahn
Daniel Heller-Roazen · Helmut Müller-Sievers
William Rasch · Joseph Vogl · Elisabeth Weber

Volume 9

Karen S. Feldman

Arts of Connection

Poetry, History, Epochality

DE GRUYTER

ISBN 978-3-11-076340-9
e-ISBN (PDF) 978-3-11-063149-4
e-ISBN (EPUB) 978-3-11-063094-7
ISSN 2195-2205

Library of Congress Control Number: 2019938425

Bibliographic information published by the Deutsche Nationalbibliothek
The Deutsche Nationalbibliothek lists this publication in the Deutsche Nationalbibliografie;
detailed bibliographic data are available on the Internet at http://dnb.dnb.de.

© 2021 Walter de Gruyter GmbH, Berlin/Boston
This volume is text- and page-identical with the hardback published in 2019.
Typesetting: Integra Software Services Pvt. Ltd.
Printing and binding: CPI books GmbH, Leck
Cover image: Elena Golikova / iStock / Getty Images Plus

www.degruyter.com

Acknowledgments

I gratefully acknowledge the opportunities for writing and research that were afforded me by the Alexander von Humboldt Foundation; the research cluster "Cultural Foundations of Europe" at the University of Konstanz, Germany; the Center of Excellence "Enlightenment – Religion – Knowledge" at the University of Halle, Germany; and the University of California, Berkeley, which provided me with sabbatical leave and other support.

Boundless thanks are due to Gilad Sharvit and James Martel for their many, many comments and responses to the various parts and iterations of this work and for their unwavering support throughout. Thanks also to Aaron Belkin, Claudia Brodsky, Judith Butler, Rüdiger Campe, Ellen Cox, Paul Davies, Eva Esslinger, Paul Fleming, Daniel Fulda, Deniz Göktürk, Hans Ulrich Gumbrecht, Robert Harrison, Anselm Haverkamp, Helmut Illbruck, Julia Ireland, Anton Kaes, Albrecht Koschorke, Sibylle Krämer, Niklaus Largier, Anja Lemke, John McCumber, Jane Newman, Shiri Sadeh-Sharvit, John H. Smith, Seán Williams, and the late Hayden White. This book is dedicated to Lulu, Toby, and Niklaus.

An earlier version of Chapter One appeared as "Unexpected Yet Connected: On Aristotle's *Poetics* and its Heterodox Receptions," in *Inventing Agency: Essays on the Literary and Philosophical Production of the Modern Subject*, eds. Claudia Brodsky and Eloy Labrada (London: Bloomsbury Academic, an imprint of Bloomsbury Publishing Plc, 2017). A shorter version of Chapter Three was published online as "On Critique and Mediality in Gottsched," online Festschrift for Prof. Sybille Krämer, Freie Universität Berlin, April 2011, http://www.geisteswissenschaften.fu-berlin.de/v/drehmomente/.

Contents

Acknowledgments —— V

Introduction: On Plot and the One-by-One —— 1

Part I: Poetry: Necessity and Plot in Aristotle and Eighteenth-Century German Criticism

1 **Unexpected Yet Connected: On Aristotle's *Poetics* and its Heterodox Receptions** —— 17
  Universals and tragic plot —— 19
  The catharsis of events —— 21
  Suspense for the sake of suspense —— 26
  The riven *Poetics* —— 32

2 **Contingency, Connection, and Possible Worlds: History and Poetry in Gottsched's *Versuch einer critischen Dichtkunst*** —— 38
  Borrowing from the possible —— 40
  Unruly *Scharfsinnigkeit* —— 44
  Fabel and possible realities —— 47
  Clear connections and the constraints of *Zusammenhang* —— 50
  Real connections in fabulated narratives —— 55

Part II: History: Aesthetic Connection in Historical Knowledge and Historical Composition

3 ***Cognitio historica* between Kant and Meier** —— 61
  Temporality and aesthetics in *cognitio historica* —— 64
  Perfect and beautiful knowledge —— 71
  Narrative knowledge and the temporality of history —— 76

4 **"On the Wings of Imagination": Wholeness and Spontaneity in Kant's Philosophy of Universal History** —— 78
  Judgment and history —— 80
  Teleology and history —— 83
  The effectivity of historical form —— 85

Freedom, spontaneity, and the infinitesimal —— 87
Continuity and the novel of history —— 90
Humboldt's mediated representation —— 92
Benjamin and Kant —— 94

5 **Not Benjamin's Ranke: On the Aesthetics of Historicism** —— 98
Form in Hegelian historicism —— 100
Ranke's constructed historicism —— 102
Not progress, not development —— 105
Ranke and ideology —— 107
Old and older historicism —— 111

Part III: Epochality: On Phenomenology's Appeals to a Disconnected Past

6 **Heidegger and the Plot of Metaphysics** —— 119
Dasein's historicity —— 121
The emplotment of inquiry —— 123
The epochal history of metaphysics —— 126
The disappearance of history —— 129
Freedom and the history of being —— 132

7 **Arendt's Epochal Phenomenology: History and the New** —— 135
Anti-Semitism in *Origins of Totalitarianism* —— 138
Freedom or narrative: Competing Kantianisms —— 143
The totalitarianism of "process" —— 147
Arendt's realist Kantianism —— 151

8 **Speaking for the Past: On *Begriffsgeschichte* and the Language of Other Epochs** —— 153
Connections to the past —— 154
Temporalities and the conditions of history —— 157
Narrative and phenomenology: Mutual interferences —— 161
Koselleck's exemplary phenomenology —— 167

Conclusion: Wholeness and its Sabotage —— 170

Bibliography —— 175

Index —— 194

Introduction: On Plot and the One-by-One

We bring to a narrative – more simply, to a story – the expectation that it form an intelligible unity, even if that unity is not immediately apparent. We expect the narrated events to fit together, to make up a meaningful whole. Structuralist, post-structuralist, and New Formalist literary criticism have investigated the ways in which plot conditions meaningful, unified narrative, with respect to fiction as well as history.[1] In this vein, the late Hayden White has shown that plot constitutes "a *pre*figurative move" that makes a set of events intelligible in the first place.[2] Historiography is therefore "fictionalized," or accomplished, using literary conventions, plot among these. Hence, literary form and historical narrative are intertwined in the representation of connections between past events.[3] While, on the one hand, emplotment can be interrogated in its relationship to historiography and the conditions of narrative intelligibility, on the other hand, in the context of literary theory plot traditionally has been understood, following Aristotle's account of μῦθος, as the arrangement of events in a literary narrative.[4] As we will see in Chapter One of this book, a multifarious literary legacy derives from Aristotle's prescriptions for the arrangement of events in a tragedy. The Aristotelian understanding of plot as an arrangement of events parallels the significance of *taxis* or *dispositio* in classical rhetoric, namely the proper arrangement of elements in the composition of a speech; in each case the overall configuration is at issue.

As an overall arrangement of events, plot configures a story as a coherent unity; the recounted events, however, must also connect to each other one by one *within* the narrative. Causality is the most obvious principle of such

---

[1] A useful brief overview is found in J. Hillis Miller, "Narrative," in *Critical Terms for Literary Study*, eds. Frank Lentricchia and Thomas McLaughlin (Chicago: University of Chicago Press 1990), 66–79.
[2] Hayden White, *Tropics of Discourse: Essays in Cultural Criticism* (Baltimore: The Johns Hopkins University Press, 1978), 1–2.
[3] See Hayden White, *Metahistory: The Historical Imagination in Nineteenth-Century Europe* (Baltimore: The Johns Hopkins University Press, 1973); White, "Introduction: Historical Fiction, Fictional History, and Historical Reality," *Rethinking History* 9, nos. 2–3 (2005): 147–157; and White, "Appendix on Narration, Narrative, Narrativization," in *The Practical Past* (Evanston: Northwestern University Press, 2014), 93–96.
[4] Peter Brooks argues that plot is not so much an overarching structure of connection but rather a dynamic configuring *operation* – that is, a temporalized, active mediation between single incidents and their unity in a narrative – and hence a matter of "plot*ting*" rather than "emplotment," which connotes a static and complete accomplishment. See Brooks, *Reading for the Plot: Design and Intention in Narrative* (New York: Vintage Books, 1984), 10, 35.

https://doi.org/10.1515/9783110631494-001

event-to-event, or "interstitial," connection of events. Certainly, events also may be represented as connected to each other in unexpected or inscrutable ways, or they may not immediately appear to be connected at all. Nonetheless, the mere presentation of events in the form of a story, even devoid of explicit reference to the connections between them, implies that the events do bear some relationship to one another. As Frank Kermode has written, "[s]equence goes nowhere without his doppelgänger or shadow, causality."[5] For the purposes of the arguments in the following chapters, it is significant that narrative in effect "insists," albeit on a formal level, that there are connections between the narrated events.[6]

This book traces, through several historical moments, the tension between the two kinds of connection described above – i.e., between the overarching arrangement or plot that holds together events from the "outside," as it were, in order to produce an intelligible whole – and the one-by-one, "interstitial" connections between events within the narrative. Events in a story can be seen as ordered according to proximate causation, which leads diachronically from one event to the next, and they can also be understood in view of the predetermining structure of the narrative as a whole. While Aristotle's descriptions of necessary connection in poetic plot are crucial, we shall also see a pinnacle of Aristotle's legacy in Chapter Two when Gottsched emphasizes the one-by-one

---

[5] Frank Kermode, "Secrets and Narrative Sequence," in *On Narrative*, ed. W.J.T. Mitchell (Chicago: University of Chicago Press, 1981), 79–97.

[6] Causality is a multifarious concept that is at the heart, for instance, of much of the philosophy of science. Proximate causation, as expressed by such causal conjunctions as "because," "as a result," and "in order to," is a quintessential characterization of necessary connection between events in legal and social-scientific discourse, see Leon Green, *Rationale of Proximate Cause* (Kansas City, MO: Vernon Law Book Company, 1927); *OED* dates the phrase "proximate cause" to 1641. In the history of philosophy the most prominent precursors to the modern notion of proximate cause are Aristotle's four causes and the notion of a prime mover (*Metaphysics* V.2–13) and Hume's attempts to distinguish constant conjunction from causality (e.g. David Hume, *A Treatise of Human Nature* [Oxford: Oxford University Press, 2000] 257–258). In the twentieth century Bertrand Russell, *Human Knowledge* (New York: Simon and Schuster, 1948), 333, describes "causal lines," and philosophy of science has developed numerous approaches to theorizing causality including probabilistic causation, counterfactuals, agency theory, and process theory. The theorization of causality exceeds disciplinary boundaries, see George Lakoff, http://blogs.berkeley.edu/2012/11/05/global-warming-systemically-caused-hurricane-sandy/, for a useful description of "systemic causation." Noel Carroll, "On the Narrative Connection," in *New Perspectives on Narrative Perspective*, eds. Willie van Peer and Seymour Chatman (Albany: SUNY Press, 2001), 21–42, provides a helpful summary of how philosophy of science dovetails with narrative theory and the philosophy of history; see also William Dray, *Philosophical Analysis and History* (New York: Greenwood Press, 1966); and J.L. Mackie, *The Cement of the Universe: A Study of Causation* (Oxford: Clarendon Press, 1974).

causal connection between events in his fabulated "history" of a land in which the depopulation resulting from war leads to famine. The ramifications of Aristotle's understanding of plot extend also to the nineteenth-century philosophy of history and to twentieth-century phenomenological appeals to bygone epochs. Whereas Kant's idea of universal history requires the idea of progress to unify the disparate events of human history into a coherent narrative, Ranke's historicism insists on the uniqueness of epochs. In the leading epochal narrative of twentieth-century phenomenology, Heidegger portrays the epochs of Western history as connected by an overarching emplotment of forgetting, but Arendt and Koselleck, in different ways, insist on profound disconnections between historical epochs.

The argument laid out in this book is as follows: these distinct but related historical moments in the history of connection demonstrate that the formal unity of narrative may compete or interfere with the representation of one-by-one connections between events in the determination of plot and vice versa. For if plot, or *Fabel* (as in Gottsched), is what makes a set of narrated events intelligible as a unity, what is the relationship of the one-by-one connections to that fabulated unity? And if history requires a coherent connection between events, where does an overarching narrative find its authority? What is more, connections between events and actions are not "things" and are not themselves "events"; they can be asserted, implied, or alluded to, but how precisely are they represented, depicted, or shown? How do poetry, history, and phenomenology accomplish the portrayal of these connections using their different theoretical and linguistic tools? In representing events as connected, both fictional and historical narratives paper over the fabulated and imaginative character of those connections, and yet the demands of poetry, history, and phenomenology each draw on different criteria and different strategies to accomplish this papering-over. Because the demands, criteria, and strategies for producing connected narratives are quite different in each case, this book does not attempt to argue for a consistent historical continuity between its moments, even though they reflect similar structural and argumentative conundrums and disconnections. To be clear, this is not a proposal for a developmental narrative for the techniques of connection from Aristotle through the twentieth century. Indeed, we will see that the problem of introducing false or at least exaggerated continuity is one that is central to the question of ideology, epochality, and historical causality. The discrepancies, however, between one-by-one connection and overarching plot structure appear in different contexts and disparate guises.

At the level of one-by-one connection between narrated events, contingency is in play, albeit perhaps latently: things might have happened otherwise; the events might have transpired differently. Hans Blumenberg's work on the concepts of reality and contingency thematizes the tension between overall narrative continuity and the representation of connection from event to event in philosophical, literary, and historical writing. Blumenberg's Kantian debts are apparent in his examination of the formal conditions of conceptuality in *Paradigms for a Metaphorology*.[7] For the question of narrative connection, however, it is more significant that in his "Lebenswelt und Technisierung unter Aspekten der Phänomenologie" [Lifeworld and technicization under aspects of phenomenology] (1963), Blumenberg focuses on "mercilessly working through the problem hiding under the quest for continuity" – that is, exposing the disjunctions that are covered over in the fabulated continuity of narrative.[8] Blumenberg also examines the novelistic arts of connection in "The Concept of Reality and the Possibility of the Novel" (1969), in which he argues that the emergence of the novel as a literary form reflects a new concept of reality as a context [*Kontext*], as opposed to earlier concepts of reality as a matter of given evidence or as what is metaphysically guaranteed.[9] Furthermore, the genre of the novel instantiates a "sphere which had not been actualized by God or nature, [...] a new reality of the possible."[10] The novel fabulates a nonactual but possible world – hence a contingent world in a new sense beyond the Aristotelian distinction between actuality and possibility.[11] For in the novel everything coheres, even if the second-order reality of the world within the novel, along with its events and the connections between them, do not actually exist.[12]

Blumenberg shows that the emergence of the novel as a literary form is significant for a history of the concept of reality. The insight that literary forms themselves are historical and historicizable phenomena is, of course, not unique to Blumenberg. To name a few others that are relevant here: Viktor Shklovsky, Jurij Tynjanov, and other Russian Formalists described the sequential development of stylistic devices; Walter Benjamin's discussion of "literary tendency" argued that literary quality depends on historical progress

---

[7] Hans Blumenberg, *Paradigms for a Metaphorology*, trans. Robert Savage (Ithaca: Cornell University Press, 2010), 4–5.
[8] Hans Blumenberg, "Lebenswelt und Technisierung unter Aspekten der Phänomenologie," *Filosofia* 14 (1963): 855–884; translation is taken from Anselm Haverkamp, "The Scandal of Metaphorology," *Telos* 158 (2012): 43.
[9] Hans Blumenberg, "The Concept of Reality and the Possibility of the Novel," in *New Perspectives in German Literary Criticism: A Collection of Essays*, eds. Richard E. Amacher and Victor Lange, trans. David Henry Wilson (Princeton: Princeton University Press, 1979), 29–48.
[10] Blumenberg, "The Concept of Reality," 40.
[11] See Rüdiger Campe, *The Game of Probability: Literature and Calculation from Pascal to Kleist*, trans. Ellwood H. Wiggins, Jr. (Stanford: Stanford University Press, 2002), 248–304.
[12] Blumenberg, "The Concept of Reality," 44.

in the techniques of literary production; and Hans Robert Jauss's manifesto for reception theory shifted the literary-critical emphasis to the audience and changes in its horizon of expectation over time.[13] "New Formalism" also belongs to this lineage, for it insists on the significance of form as a historical, political, and social phenomenon while granting form its own effectivity within these realms.[14] Blumenberg, however, most pointedly conveys the uncomfortable convergence between the formal constraints of narrative and the representations of worldly contingency and possibility. This awkward convergence, as we will see, characterizes not only the novel but also the historical account and the phenomenological reference to historical epochs. What is more, the points of convergence and of awkwardness shift as the historical context and genre of text change. In the course of the development of universal history, for instance, the representation of history as a unified form is at stake; in German historicism, the specificity of each moment demands a different narrative form.

Blumenberg's concern for the "problem hiding under the quest for continuity" illustrates the following problem, whose different dimensions are illuminated in the three parts of this book: insofar as narratives reflect worldly contingency, the continuity of events cannot be portrayed as entirely necessary within the plotline of the narrative.[15] Nonetheless, the representation of contingent connections between individual events potentially interferes with or subverts the narrative's formal unity. Especially where human action is in play, the sense that things might have happened otherwise and thus that the represented events are contingent, haunts even the most deterministic of narratives. This tension between the overarching structure of plot and the one-by-one connection of events is visible, for instance, when historical events are portrayed in terms of contingency and human freedom, and yet this portrayal is accomplished within an organized narrative whose overarching unity implicitly denies that very contingency and freedom. Such seemingly contingent connections stand out against a backdrop of confidence in the unity of the narrative trajectory, which is the

---

13 See Jurij Tynjanov, "On Literary Evolution," in *Readings in Russian Poetics: Formalist and Structuralist Views*, eds. Ladislav Matejka and Krystyna Pomorska (Champaign: Dalkey Archive Press, 2002), 66–78; Jurij Striedter, *Literary Structure, Evolution and Value*, trans. Matthew Gurewitch (Cambridge: Harvard University Press, 1989); Walter Benjamin, "The Author as Producer," in *Selected Writings*, vol. 2, no. 2, eds. Michael Jennings, Howard Eiland, and Gary Smith (Cambridge: Harvard University Press, 2005), 768–782; Hans Robert Jauss, "Literary History as a Challenge to Literary Theory," in *Toward an Aesthetic of Reception*, trans. Timothy Bahti (Minneapolis: University of Minnesota Press, 1982), 3–45.
14 See Marjorie Levinson, "What is the New Formalism?," *PMLA* 122, no. 2 (2007): 558–569.
15 Anselm Haverkamp, "The Scandal of Metaphorology," *Telos* 158 (2012): 39.

promise of storytelling. How does that larger unity limit the waywardness or even the freedom of the human beings whose individual moments and connections are told in a story? This book argues that such one-by-one, contingent connections of narrated events both shape and come into tension with the very form of unitary narrative. What is more, the very portrayal of human freedom itself, as in the phenomenological enterprises of the twentieth century, seems to depend on a characterization of history as defined by distinct epochs.

In the chapters to follow, we will investigate instances in which the theorization of formal narrative unity indirectly, and in disparate ways, imposes a constraint upon the contingency and waywardness of individual moments or sequences of events that appear within a narrative of human action.[16] We will observe how in different eras and in the contexts of different demands of genre, prescriptions for producing a narrative – whether historiographical, literary, or phenomenological – must propose a method both for connecting individual events to each other one by one and for producing an overarching unity. With regard to the self-conscious prescriptions for poetic writing, Aristotle is key and Gottsched represents the eighteenth-century German hope for aesthetic prescription and the connections of poetic thinking. In the development of modern historiography, Kant and Ranke grapple with the problem of continuity and necessity in the representation of historical events. In the twentieth century, phenomenology far exceeds a synchronic plane of analysis when it attempts to define modernity and incorporates historical narrative into its attempts to differentiate fundamental human phenomena from historicizable variations in human existence. For these reasons, much of this inquiry builds on theorizations of plot and hence on narratology, for with plot it is precisely the overall structure and unfolding of narrative that is at stake. We will expose tensions that appear between principles, artistic methods, and the extra-textual, structural logics of plot or trajectory on the one hand and the interstitial or one-by-one connection from event to event on the other hand. For individual moments and connections are the stuff of any narrative but also harbor resistance to its larger continuity. The possibility of such intranarrative

---

16 See Albrecht Koschorke, *Wahrheit und Erfindung: Grundzüge einer Allgemeinen Erzähltheorie* (Frankfurt am Main: Fischer, 2012), 27–110, for a comprehensive treatment of the elementary operations of narrative. For a narrower approach that focuses on the future perfect tense in a deconstructive mode, see Mark Currie, *The Unexpected: Narrative Temporality and the Philosophy of Surprise* (Edinburgh: Edinburgh University Press, 2013); Kasia Jaszczolt, *Representing Time: An Essay on Temporality as Modality* (Oxford: Oxford University Press, 2009), considers the representation of time and possibility from the perspective of philosophy of language and linguistics.

resistance raises such questions as how to represent the connection between proximal events, how the imagination produces connections and unity according to or in violation of rules of composition, how readers or viewers are moved by a story, and how a whole can represent each of its parts as something other than entirely necessary.

This inquiry will therefore scrutinize prescriptions for connecting events – what we could also call normative representations of connection – for literary and historical, i.e., fictional and nonfictional, narratives. When we examine prescriptions for writing historical, literary, or philosophical narratives, we see how the requirement for one-by-one connections from event to event competes and interferes with the requirement for an overall organization of the narrative. In prescribing principles for producing narratives, precisely this tension between the individual events of the narrative and the overarching unity of the narrative is at issue.[17] Such prescriptions involve rules for the relationship of the parts and the whole of a narrative, and yet they also run into limitations regarding how the juxtaposition of events can challenge the presumption of formal unity that defines a story. The various modes of storytelling that are to be found in classical and neoclassical poetry, in the development of modern historiography, and in the phenomenological accounts of modernity all inhabit the tensions that surround the prescriptions for narrative connection.

This book will also argue that there is a peculiar problem posed to narrative continuity by the representation of human action and spontaneity, for spontaneity marks a particular kind of departure from necessary connection to antecedent events. Whether in literature, history, or philosophy, as we will see, the portrayal of human freedom as a spontaneous force entails the representation of profound contingency within a sequence of connected events. Nonetheless, the emergence of unconditioned spontaneity must square with the formal unity of the narrative. Hence, in philosophy since Kant, the question emerges as to how freedom can be represented in historical narratives and by which conventions. When freedom is defined in the liberal tradition of the Enlightenment as the sovereign capacity of an individual to act, its proper mimesis in the narrative poses a problem, for the historical narrative as an integral structure constrains what the historical actors who people its world are able to do. People or characters within a narrative, in other words, cannot be "truly" represented as free because they do, after all, inhabit the narrative and are subject to the demands of formal unity and

---

17 Paul Fleming, "The Perfect Story: Anecdote and Exemplarity in Linnaeus and Blumenberg," *Thesis Eleven* 104, no. 1 (2011): 74–75, describes in a similar fashion the discrete and contingent character of anecdotes.

coherence. The framing of the narrative as a story thus controverts the absolute sovereignty of its characters. This means that the politics of history are constrained not only by the stories that are told, but also by the very form of telling.[18] So how are human freedom and, more abstractly, possibility as such represented narratively if narrative formal unity seems to foreclose the representation of genuine contingency and spontaneity? How does the structural, emplotted aspect of storytelling interfere with the contingency that is only imperfectly represented within a story, in particular where human action and freedom are constrained by the very narratives that should portray them?

The following chapters respond to these questions by surveying specific instances in which plot or the overarching narrative arc comes into tension with prescriptions for the representation of contingency. Such contingency may take the form of freedom, spontaneity, wayward connectibility, and ungovernable imagination. In the context of poetry, plot involves an artistic fabulation; in history, there is a demand for fidelity to worldly events; in phenomenology, the description of modernity delineates specific characteristics that define an epoch. We will see that the unity of plot and the intelligibility of narrative require connections between individual events but cannot represent the connections between events and epochs directly, just as causality itself cannot be represented or discerned directly but instead only inferred. The chapters that follow therefore deal, broadly speaking, with the confluence of literary techniques, historiographical prescriptions, and philosophical demands for producing narrative continuity – i.e., the formal conditions of representing connections between events – in the context of poetry, the philosophy of history, and twentieth-century phenomenology. They will thematize the hazardous art of narrative connection, in particular the literary and rhetorical devices that both create connections between events and also disrupt them. Each chapter explores a punctual moment in a lengthy historical trajectory, beginning with Aristotle's formulations regarding plot; via Gottsched and the critical differentiation of literature, history, and philosophy in the eighteenth century; through the appearance of the philosophy of history from Kant to Ranke; to twentieth-century invocations of history in phenomenology. It is not the goal in these pages to provide a comprehensive history of thought on these topics. Rather, each of the

---

18 Gary Saul Morson, *Narrative and Freedom: The Shadows of Time* (New Haven: Yale University Press, 1994), argues that both historical and literary narratives that depict an inexorable movement toward a particular end, mainly by use of such techniques as foreshadowing, in effect offer readers a training in unfreedom. Narratives that emphasize contingency at each moment and a non-deterministic sequence of events, in contrast, maintain our sense for the contingency of the present and for the possibilities of human freedom.

thinkers treated here epitomizes a particular key moment in the shifting relationships between poetry, history, and philosophy, in which the question of how events are connected pushes to its limit the genre of discourse in question. Indeed, when we look at prescriptions for composing narratives, we may observe attempts to contain contingency within plot and overall form that nonetheless also seek to preserve and represent contingency within the sequence of narrated events. It is precisely at this junction that narratives shuttle between the second-order reality within their pages and extra-narrative hopes or fears regarding worldly contingency and human freedom.

Part I deals with the variegated reception of Aristotle's prescriptions in the *Poetics* for connecting tragic events. Chapter One centers on Aristotle's emphasis within his account of tragedy on the centrality of plot, for the emplotted trajectory of the tragic action as a whole demands a strictly causal relationship between key individual events. For the legacy of literary criticism, Aristotle provides the most influential and succinct expression of the requirement that events be represented as connected to each other. Nonetheless, there is much dispute regarding how to interpret Aristotle's prescriptions for connection – e.g., in terms of precisely how the individual tragic events should be connected, the effect of this connection on the audience, and the very purpose of the tragedy that depends on this connection. Aristotle's *Poetics* is also, of course, a cornerstone when it comes to canonical claims for the value of literature. In traditional readings, the *Poetics* is understood to be saying that the value of tragedy – and by extension the value of literature – lies in the following: it offers a catharsis of fear and pity, and insight into our own finitude, most particularly by way of a climactic plot that turns on the unexpected conjunction of events. Chapter One contrasts these canonical claims about the value of tragedy and literature as such with several non-canonical, even heterodox, interpretations of how and why the tragic events must be so tightly connected. It highlights widely divergent views of the *Poetics* in interpretations ranging from Lessing to twentieth-century philosophy in order to unsettle canonical understandings of Aristotelian plot.

The *Poetics* was also central in the emergence of critique proper in the eighteenth century, as the operations of poetry, historiography, and philosophy came to be differentiated from one another. Within the German tradition, Gottsched's 1730 treatise *Versuch einer critischen Dichtkunst* (Attempt at a critical art of poetry) highlights a key moment in the emergence of distinct disciplines.[19] Chapter Two looks at the exemplary, and for non-German readers largely unknown, case

---

[19] See Rüdiger Campe, *Affekt und Ausdruck: Zur Umwandlung der literarischen Rede im 17. und 18. Jahrhundert* (Tübingen: Niemeyer, 1990), 3–8; 39–53.

of Gottsched's eighteenth-century uptake of Aristotle's poetic prescriptions. Gottsched's work offers an important angle on the reception of Aristotelian thought. He formulates rules derived from Aristotle in order to attempt to govern the promiscuous association of thoughts in poetic writing. The chapter will show, however, that the ungovernable connectibility of thought raises difficulties when it comes to regulating the principles of connection between narrated events in historical and poetic writing. For thoughts can connect to other thoughts ungovernably and yet both poetic and historical narratives require, in an Aristotelian vein, a principle of connection and a governing unity. Hence with regard to Gottsched's *Dichtkunst*, Chapter Two considers the connections of events in light of his reception of Aristotle's *Poetics*; the role of *Witz* [ingenuity], *Scharfsinnigkeit* [acuity], and imagination; and the definition of *Fabel* [plot or fable]. Gottsched claims that there is a difference between the connection of events in poetic narratives and the connection of events in historical narratives. But Gottsched's attempt to illustrate this difference between the principles of connection in poetry and in history-writing, as the chapter will show, confound his own prescriptions. In particular, his treatment of *Fabel* indicates that whether historical or poetic, narrative is in each case a hybrid of actuality and possibility, worldliness and subjectivity, promiscuous associability and rules.

Part II focuses on the significance of artful connection in the development of the Kantian and Rankean philosophies of history. Whereas poetic continuity obviously involves the writer's craft in order to create a compelling plot, Chapter Three investigates a subtle eighteenth-century modification in the now-obscure notion of *cognitio historica*, or historical knowledge, that marks the introduction of diachronic narrative into the concept of history itself. In the German tradition before Kant, "historical knowledge" referred to empirical knowledge rather than to anything historical in the modern sense. Kant's early lectures on logic, based on a handbook by Georg F. Meier, explain historical knowledge instead as knowledge of the past. In that context, Kant's descriptions of historical knowledge invoke examples of diachronic narratives of connected events rather than synchronic descriptions of static elements. Kant's lectures thereby illuminate a move toward a temporalized understanding of history and toward a view of the past as a narratable series of connected events. This largely unremarked shift in the notion of historical knowledge, Chapter Three will argue, suggests that Kant developed a temporalized and narrativized notion of history.

Chapter Four takes up a more familiar topic in Kant, namely his reflections on the connection of events in the project of universal history. Kant, and Hegel after him, thematizes the apparent incompatibility between the narration of worldly events on the one hand and the philosophical occupation with concepts and reason on the other. Here, the question of how events are connected to each other is seen as an explicitly philosophical one, insofar as rational connections are imposed upon worldly happenings in order to produce a historical account that matches the philosophical demands of reason. Nonetheless, freedom for Kant is understood as a matter of self-governing, self-causing, spontaneous action, i.e., action not determined by what comes before. Freedom is thus "anarchic" or unprecedented; it occurs as a breach with respect to the events that preceded it, i.e., an interruption of causality. And yet, how can a unitary story represent human freedom if freedom is precisely a break from causal connection? In the context of history-writing, the problem becomes a narrative one: what would it mean to describe human actions as not determined by what came before and therefore as contingent?[20] The consideration of the structural unity of historical narrative poses a problem to the representation of freedom. Does Kant, in other words, face a historiographical problem with how freedom is to be represented in a unitary narrative, when in his practical philosophy freedom functions as a *dis*connection from any foregoing events and causalities? In the wake of Kantian aesthetics there emerged the question – and various solutions – regarding how to represent events as causally connected one by one in historical contexts while also depicting them as part of a unitary human history by means of concepts, above all "progress," deployed as plotlike devices.

Chapter Five takes up the reception of Ranke and German historicism, in which Kantian universal history and its demand for a unitary presentation is rejected in favor of other philosophical and aesthetic principles. Its claim is that if we take Walter Benjamin at his word in his description of a linear, positivist tradition of German historicism, we neglect the aesthetic, disruptive, and even anti-ideological elements that can be found in Ranke's rejection of universal history and in his prescriptions for aesthetic principles of historiography. Chapter Five shows that Ranke's prescriptions for historiography, for instance his commentary on Italian Renaissance historian Francesco Guicciardini, speak against Benjamin's characterization in a variety of ways. What is more, Ranke's rejection of historiographical predecessors may itself be seen as a *topos*, a historical misprision.[21] For instance, just as Ranke rejects the techniques of earlier historians, Johann Gustav Droysen in his turn rejects Rankean historicism on both aesthetic and historical grounds. The self-conscious development of the art of connection in historiographical

---

20 See Campe, *The Game of Probability*, 166–171, for more explanation of the development of a notion of contingent causality.
21 The formulation "historical misprision" is here a variation on the "poetic misprision" made famous in Harold Bloom, *The Anxiety of Influence: A Theory of Poetry* (Oxford: Oxford University Press, 1997), 19–38, 83.

representation seems to demand a gesture of *dis*connection from historiographical predecessors and their artful procedures.

Part III explores the adjuvant role of plot in the historical ambitions of twentieth-century phenomenology as found in the epochal narratives deployed by Heidegger, Arendt, and Koselleck. Phenomenology encompasses numerous strands of inquiry and bears the marks especially of Hegel and Husserl. The present context attends to the epochal stories told by mid-twentieth-century phenomenology in attempting to account for modernity itself. Each chapter lays out a distinct attempt on the part of post-Husserlian phenomenology to reclaim history for philosophy in ways that complicate the relationship between overarching narrative trajectory and one-by-one connections of events. Through the 1920s, Heidegger used the term *Geschichtlichkeit* [historicity] to describe our human way of being-in-the-world. Heidegger's procedures for investigating this historicity follow an arc or what we could call a plot structure – i.e., dismantling a series of everyday conceptions and uncovering a "foundational" event of disclosure and concealment. For Heidegger, historicity also comes to designate the epochal history of metaphysics. Chapter Six argues that Heidegger's epochal narrative of the forgetting of being recapitulates and allegorizes, in its emplotted form, precisely the stepwise procedure of Heidegger's account of Dasein's historicity. Chapter Six also suggests that this procedure is typical across Heidegger's corpus: a discovery of a forgotten or eclipsed eventlike disclosure is recounted in the history of metaphysics but also represents the path of Heidegger's general mode of phenomenological inquiry, including his essays "On the Essence of Truth" and "The Origin of the Work of Art." The conformity of his epochal story with the "plot" that structures his own investigations represents a formal continuity between his phenomenological inquiries and his epochal narratives.

Arendt's *Origins of Totalitarianism,* in contrast to Heidegger's long sweep of historical narrative, takes up a particular moment in history in order to illuminate the conditions of politics, *Mitsein* [being-with-others], and freedom. Chapter Seven demonstrates, however, that Arendt's historical approach to the investigation of totalitarianism interferes with her phenomenological goal. In particular, her claims for newness and the significance of the unprecedented is undermined by the inadvertent elements of emplotment that structure her massive historical study. Within particular textual moments and turns of phrase, a tension appears between Arendt's analysis of the historical origins of totalitarianism (as promised by the book's title) and her phenomenological description of the events and elements that mark totalitarianism's emergence. The chapter argues that insofar as Arendt makes use of historical material, her claim for the radical contingency and newness of the totalitarianism she recounts is compromised. Taken together, Chapter Six and Chapter

Seven show that Heidegger and Arendt each make epochal claims about history – claims about how events are connected over particular stretches of time and disconnected from other stretches of time – in ways that complicate their preponderant concerns for a quasi-original disclosure of possibility (Heidegger) and human freedom and spontaneity (Arendt). Heidegger composes an epochal history of being and Arendt offers an epochal analysis of totalitarianism, but for both, the element of plot intrinsic to historical narrative works against the task of illuminating possibility, freedom, and spontaneity, for these are unsuited to the tight connection of events entailed by plot.

Chapter Eight investigates Reinhart Koselleck's *Begriffsgeschichte*, or conceptual history, of the concept of history itself. Koselleck claims that in premodern history-writing, stories about the past served not as information about the past, nor as a way causally to connect the events of the present to the past, but as exempla offering lessons to the present within a cyclical temporality.[22] There was no question of speaking *for* the past because there was no deep disconnection *from* the past in the first place. Only with the advent of modern historiography in the late eighteenth century, so Koselleck's story goes, was the past seen as essentially disconnected from the present, as something historiography could "speak for," which therefore demanded styles of writing that would produce a connection to that past. In defining a new sense of historicity after 1800, Koselleck produces an epochal history. That epochal narrative also serves as the foundation for a phenomenological study of the structure of historical time and a quasi-transcendental investigation of conditions of possible histories. Koselleck's structural and quasi-transcendental inquiries nonetheless operate in some tension with the *begriffsgeschichtliche* narrative of epochs. In fact, as the chapter will show, Koselleck's phenomenology of historical time undermines itself in key respects, each of which are related to the thorny problem of how to represent connections between events.

The authors treated in these chapters illuminate a variety of possibilities for representing connection – whether in terms of imagination, reason, causality or plot – and for the deployment of formal and figural elements that come into play in each case. In each of these moments at the nexus of history, philosophy, and literature, the various claims for how connections are made or should be made reflect otherwise unthematized dilemmas regarding how to represent contingency, possibility, and freedom at the level of form. The ungovernable ramifications of connection and disconnection provide a bridge, albeit an unstable

---

[22] Reinhart Koselleck, *Futures Past: On the Semantics of Historical Time*, trans. Keith Tribe (Cambridge: MIT Press, 1985).

one, when it comes to the supposed divide within the context of "theory" between formalism and historicism. In other words, narrative as such at the formal level has an effectivity that belongs to a particular historical moment without being reducible to it. The tensions between interstitial, one-by-one connection and overarching plot are different in each case, but the goal is to show that the disjunction between them both produces and threatens to thwart the dynamic unfolding of narratives.

Part I: **Poetry: Necessity and Plot in Aristotle and Eighteenth-Century German Criticism**

1 Unexpected Yet Connected: On Aristotle's *Poetics* and its Heterodox Receptions

The prescriptions in Aristotle's *Poetics* for the connection between tragic events shape the canon of literary criticism, albeit in disparate ways. The *Poetics* famously states that of all the components of tragedy, the arrangement of events [τῶν πραγμάτων σύστασις] is the most important (1450a15).[1] That arrangement, the plot [μῦθος], structures the tragic incidents so that they form a cohesive unity. Aristotle notes that the intrinsic integrity of tragedy stands in contrast to the adventitious composite that is history. In a historical account, the recounted events need be connected only by virtue of occupying the same span of time:

> [Tragedy's] structures should not be like histories, which require an exposition not of a single action but of a single period, with all the events (in their contingent [ἀνάγκη οὐχί] relationships) that happened to one person or more during it. (1459a20–22)[2]

Unlike history, tragedy involves the exposition of an action that is unified and complete in itself (1459a16). The plot or arrangement of incidents is what ensures, from the inside, as it were, that they compose a *whole* incident, a *complete* event. The particularities of plot form one side of the dynamic that stands at the center of this book: the oscillation between the operations of overarching structural emplotment and the one-by-one connections of events.

What precisely is it, however, that renders a series of events a unitary and complete action? Aristotle describes historical narratives as portraying events whose relationships to each other are only "contingent," "casual," or, translated more literally, "not necessary." The contours of those relationships are defined by a central personage or the designated limits of certain periods of time, rather than by any intrinsic connection. The historian's job therefore is to report events "whether or not the events exhibit explanatory coherence."[3]

---

[1] Aristotle, *Poetics*, trans. Stephen Halliwell (Cambridge: Harvard University Press, 1995). All subsequent translations are taken from this edition, except where otherwise indicated.

[2] W. Rhys Roberts's translation suggests that the events in a historical narrative may be merely "casually," rather than "contingently," related to each other; see Aristotle, *Poetics*, trans. W. Rhys Roberts (London: Heinemann, 1927), http://data.perseus.org/citations/urn:cts:greekLit:tlg0086.tlg034.perseus-eng1: 1459a.

[3] J. M. Armstrong, "Aristotle on the Philosophical Nature of Poetry," *The Classical Quarterly* 48 (1998): 447.

The events of a history need display no constitutive unity, no principle of essential connection, no intrinsic coherence. Tragedy, in contrast, should not be structured in this contingent fashion. It should instead portray necessarily related incidents that cohere and thereby form a unity. The tragic incidents must be tightly connected, each one to the next. Aristotle writes that the moments of recognition and reversal that are a hallmark of tragedy "should emerge from the very structure of the plot, so that they ensue from the preceding events by necessity or probability" (1452a18). No matter how sad or regrettable they may be, accidental and unrelated events are not tragic; the necessary connection between the incidents is what is definitive of tragic plot.

Aristotle's prescriptions for complete action, plot, and intrinsically connected events have been assigned different orders of significance in the history of literary theory. Canonical views of the value and purpose of tragedy claim variously that the requisite tragic elements in their proper connection purge the emotions of fear and pity, render us more sensitive to suffering, give us insight into human finitude, or teach universal truths. These views of tragedy have shaped a range of views of the purpose of literature in general and its value for knowledge and human existence. It is the goal of this chapter, however, to investigate several lower-stakes interpretations of the significance of plot in the *Poetics*. In their divergences from the above-mentioned "high stakes" interpretations, they illuminate heterodox interpretative possibilities for understanding the *Poetics*' prescriptions for the connection of tragic events. They also undermine canonical philosophical, humanistic, and aesthetic values that are said to derive from Aristotle's text. While the interpretations of the *Poetics* that we will consider deflate some of the loftier and better-known claims made about the purpose and value of tragedy, it is not the goal of this chapter to adjudicate their philological plausibility and fidelity to Aristotle. Rather, the very fact of such interpretative disparities is the interest here. In this approach, our argument follows Peter Szondi's assertion that "the history of modern poetics is the history of [the *Poetics*'] reception and influence," including its "adoption, expansion, and systematization, as well as misunderstanding and critique."[4] Whether or not they are believed to be faithful to Aristotle or are accepted by humanists in general, the following deflationary interpretations of plot and of the principle of connection between events in the *Poetics* produce implications for literary theory as a whole, unsettling claims about the value of literature and the arc of literary history.

---

[4] Peter Szondi, *An Essay on the Tragic*, trans. Paul Fleming (Stanford: Stanford University Press, 2002), 1.

Universals and tragic plot

If we want to make large-scale claims for literature, and for the difference between literature and historiography, the *Poetics*' reference to poetry and universals is an attractive starting point. Aristotle writes that "Poetry [...] is more philosophical and more elevated than history, for poetry relates more of the universal [τὰ καθόλου], while history relates particulars" (1451b1–3). Jonathan Lear articulates a canonical position on these lines, namely that "[t]he universality Aristotle has in mind when he talks about the universality of poetry [...] is aiming at the universality of the human condition."[5] Similarly, Aristotle's references to universals have been widely interpreted to mean that tragedy deals with broader, wider-reaching truths than other kinds of writing. William Faulkner expresses as much in his speech upon receiving the Nobel Prize for Literature in 1950, which in the context of what he calls "[o]ur tragedy today" demanded that writers include in their stories "the universal truths lacking which any story is ephemeral and doomed."[6]

We find a much more modest philosophical yield from tragedy in J.M. Armstrong's little-known interpretation of universals in the *Poetics*. Armstrong rejects the notion that the universals with which tragedy is engaged are universal truths or the human condition.[7] Instead, Armstrong translates *Poetics* 1451b8–11 as explaining that

> [a] universal is the sorts [sic] of thing that a certain sort of person happens to say or do according to likelihood or necessity, which is what poetry aims at, although names [of characters] are added.[8]

Armstrong thus interprets universals simply as types of actions, i.e., as generic plot structures. In this interpretation, if there is learning from tragedy, what is learned is not a universal truth. More modestly, tragedy teaches us about the relation between human characteristics "and the actions which those traits tend to produce."[9] In other words, tragedy is a matter of portraying connections

---

5 Jonathan Lear "Katharsis," in *Essays on Aristotle's Poetics*, ed. Amélie Rorty (Princeton: Princeton University Press, 1992), 326.
6 William Faulkner, "Banquet Speech," *Nobelprize.org*. Nobel Media AB 2014. Web. 5 Jun 2017. http://www.nobelprize.org/nobel_prizes/literature/laureates/1949/faulkner-speech.html
7 Armstrong, "Philosophical Nature of Poetry," 449–450.
8 Armstrong, "Philosophical Nature of Poetry," 451–452; where Armstrong is cited in a note, the translation is his own.
9 Armstrong, "Philosophical Nature of Poetry," 455.

between events, actions, and the type of people who perform such actions. This is "universal" in that tragedy abstracts from particulars to types and to the connections that follow accordingly.

As further evidence in support of his interpretation, Armstrong examines Aristotle's brief attempt to clarify what he means by universals and particulars. Aristotle contrasts the universal to a particular set of incidents, citing as an example of the latter "what Alcibiades did and what he experienced" (1451b11).[10] Armstrong rejects a standard interpretation in which this passage is read as claiming that universals refers to human beings in general, as opposed to particular individuals. Armstrong argues that the particular to which Aristotle refers here is not Alcibiades but instead the particular *events and incidents* that Alcibiades experienced – an interpretation of the passage that some translations, including Halliwell's, support.[11] Armstrong's interpretation of "universal," however, ascribes to it no large-scale significance. He takes it to refer merely to a genre or type of action when it comes to forming a generic plotline to be peopled with particular characters, deeds, and incidents.[12] Accordingly, tragedy's universal element does not refer to the conditions of human existence per se. Rather, universality is a matter of the generic structure of tragedy and of the connections between actions and the characters most likely to perform such actions. A "type of action" is, in this reading, "universal" because it involves a generic structure that can be repeated in many contexts with the insertion of different particular incidents and characters – as long as those characters are of types likely to perform the requisite actions.

If, as Armstrong claims, the universal is merely the generic plotline with likely characters, rather than a matter of a higher truth, then why does Aristotle declare that poetry is "more philosophical" than history – a statement that has lent credence to many claims for the edifying nature of poetry? In Armstrong's interpretation, it is not the tragedy that is philosophical, nor does it produce philosophical effects on the audience. Aristotle means that what is philosophical is the poet's understanding, by means of which the poet "discern[s] the structure" of the appropriate type of action.[13] The poet's task of constructing the work is therefore "more philosophical" because it "involv[es] more understanding," namely of different generic plots and characters and the appropriate

---

10 Armstrong, "Philosophical Nature of Poetry," 450.
11 Armstrong, "Philosophical Nature of Poetry," 449.
12 O.B. Hardison, Jr., "Commentary on Aristotle's *Poetics*," in *Aristotle's Poetics: A Translation and Commentary for Students of Literature* (Tallahassee: Florida State University Press, 1981), 118–120, offers a summary of didactic, theological, and other views of universals in Aristotle.
13 Armstrong, "Philosophical Nature of Poetry," 447

connections to make among them.[14] Aristotle's references to the difference between knowledge and skill vs. mere experience in the *Metaphysics* are significant here. In the *Metaphysics*, Aristotle writes that knowledge and skill involve an understanding that results from encounters of similar particulars (981a5–6).[15] In a similar way, for Armstrong the poet must understand the universal – the type of action, the type of character, and the generic plot that arranges the events – in order properly to compose his tragedy. The "universality" of the generic action takes proper connection as the crux of tragedy. The poet is philosophical because he understands how such generic connections should be made. This understanding of universals promotes no insight into fate, inexorability, or tragic flaw. It instead concerns a clarity about the appropriate generic structure of incidents and about the proper connections between events, characters, and deeds. Armstrong's interpretation of Aristotle's "universals" elevates the significance of the overarching structure of emplotment but thereby deflates a dominant reception of the *Poetics* in which Aristotle is seen to claim that tragedy teaches a universal human truth. The structural and characterological discernment of the poet, rather than tragedy itself, is what is "philosophical" because the principle of connection between events and characters is central.

The catharsis of events

Catharsis is perhaps the best-known single term from the *Poetics*, although it also appears in other texts of Greek antiquity. There are, however, multifarious understandings of catharsis in the reception of the *Poetics*, each of which bears different implications for the value of literature. This chapter will show that even this term that seems to put emotions at the center of tragedy, and which has been taken for the last 250 years as a humanistic and pedagogical principle of literature per se, can instead be read, in deflationary fashion, in terms of the mere existence of connections between the tragic events. There are, in other words, several ways to interpret the *Poetics* in terms of emplotted connection, without regard to emotions, the audience, and the high-stakes implications for the humanistic enterprise that have been drawn from Aristotle's text.

Most translations of *Poetics* 1449b have rested on interpretations of catharsis as the purgation of emotions belonging to the audience, for instance: "Tragedy, then, is mimesis of an action [...] and through pity and fear accomplishing the

---

14 Armstrong, "Philosophical Nature of Poetry," 448.
15 Armstrong, "Philosophical Nature of Poetry," 448.

catharsis of such emotions [περαίνουσα τὴν τῶν τοιούτων παθημάτων κάθαρσιν]" (1449b23–28). In this interpretation, the tragic action evokes fear and pity in the audience, culminating in a climactic release of that fear and pity. This "purgation" or "purification" of fear and pity has been seen as valuable, healthy, or as rectifying a pathology.[16] The very idea of such a homeopathic evocation and purgation of these emotions implies that they need reduction or balancing. Tragic catharsis has also been interpreted as cleansing a moral pollution in the world or in the audience.[17] (Of course, the word "catharsis" has had an illustrious career in psychology, most notably with Freud but also in contemporary abreaction therapy, in which it is used to treat post-traumatic stress disorder, among other ailments.[18])

These interpretations of catharsis downplay its relationship to the connection of events. Instead, they emphasize that literature evokes especially important emotions in an especially valuable way. For instance, our sense for the commonality of all other human beings is said to develop in viewing tragedy and experiencing the purgation of fear and pity. This characterization belongs to a significant tradition of interpretation in which the *Poetics* is seen not only to offer lessons in morality and human finitude but also to serve as a valuable means of sentimental education, making us more sympathetic to the plight of others. Sympathy turns out to be a complicated emotion, however. Whether sympathy should be understood as a matter of emotion or rationality was a central question in eighteenth-century German drama. The combination of fear and pity turns out to be a key to this question. For Lessing pity is, by its very nature, fearful because it involves a rational recognition of sameness between the one who pities and the one who is pitied. We feel pity for the hero and fear for ourselves. Lessing insists in §75 of the *Hamburg Dramaturgy*, against Moses Mendelssohn, that the fear to which Aristotle refers is not fear for the hero but rather

> es ist die Furcht, welche aus unserer Aehnlichkeit mit der leidenden Person fuer uns selbst entspringt; es ist die Furcht, dass die Unglueksfaelle, die wir ueber diese verhaenget sehen, uns selbst treffen koennen; es ist die Furcht, dass wir der bemitleidete Gegenstand selbst werden koennen. Mit einem Worte: diese Furcht ist das auf uns selbst bezogene Mitleid.

---

[16] See Hardison, "Commentary," 132–137.
[17] See Leon Golden, "The Purgation Theory of Catharsis," *The Journal of Aesthetics and Art Criticism* 31, no. 4 (1973): 473–479.
[18] S.W. Jackson, "Catharsis and Abreaction in the History of Psychological Healing,"*Psychiatric Clinics of North America* 17, no. 3 (1994): 471–491.

It is the fear for ourselves, which derives from our similarity with the suffering person; it is the fear that misfortune that we see befall others could also befall us; it is the fear that we could become the object of pity. In a word, this fear is pity directed at ourselves.[19]

As we see, for Lessing fearful pity is a matter of combined emotions, but for the Enlightenment tradition what is more important is that fear and pity together constitute a combination of emotion per se with rationality. Tragedy, in other words, calls on a rational recognition that is already present *in the emotions themselves* of our shared human finitude and proneness to disaster.

According to Lessing, tragedy thus can be understood as oriented toward producing effects on the audience that combine rationality and emotion. In this context, Lessing laments the German translations of Aristotle in which φόβος is translated *Schrecken* [horror] instead of *Furcht* [fear]. He in effect suggests, without naming the rhetorical figure, that the phrase "fear and pity" should be understood as a hendiadys. Hendiadys is a figure in which two nouns are employed to express a single idea, rather than a noun and an adjective – e.g., "rain and weather" instead of "rainy weather."[20] Lessing claims that catharsis evokes not "fear and pity" but "fear*ful* pity," *mitleidiges Schrecken*, insofar as pity by definition must contain an element of fear in the face of uncertainty regarding one's own fate.[21] Tragedy's evocation of fear and pity serves to cultivate empathy in the audience, for it mixes pity for the hero with fear for oneself amid a rational recognition of our likeness.[22] Hence in this reading rationality, fear, and pity are closely connected: we pity the tragic hero but are fearful insofar as we rationally recognize that we could be similarly met by such an unpleasant fate. Because we identify with the hero and hence feel pity for him, we know that we should fear for ourselves, all of which makes us more sensitive, empathic, humane, and increases our sense of fellowship with all human beings. Recognition of our shared humanity, our proneness to inadvertent transgressions, our shared vulnerability to fate – all of these shared human foibles are depicted in the tragedy.[23] The formal, aesthetic, or philosophical

---

19 G.E. Lessing, *Hamburgische Dramaturgie*, in *Werke*, vol. 4, ed. Herbert G. Göpfert (Munich: Hanser, 1979), 267; *Hamburg Dramaturgy*, trans. Helen Zimmern (New York: Dover, 1962).
20 "Silva Rhetoricae/The Forest of Rhetoric" (website), rhetoric.byu.edu. For more on hendiadys see my "On Vitality, Figurality, and Orality in Hannah Arendt," in *Thinking Allegory Otherwise*, ed. Brenda Machowsky, 237–248 (Stanford: Stanford University Press, 2013).
21 Lessing, *Hamburgische Dramaturgie*, 265.
22 Lessing, *Hamburgische Dramaturgie*, §§74–78, 264–277.
23 See Claudia Brodsky, "Lessing and the Drama of the Theory of Tragedy," *MLN* 98, no. 3 (1983): 426–453, on Aristotle and Lessing. Brodsky differentiates between earlier *Handlungs-* or *Schicksalstragödien* and later *Charaktertragödien*. In early Lessing, she explains, the

principles that connect the events to one another serve to evoke the crucial fearful pity that combines emotion and rationality.

Similar claims for the value of tragedy in cultivating empathy for others have been extended to literature as a whole in twentieth-century philosophy. Richard Rorty emphasizes the value of certain works of literature in helping us to become "less cruel."[24] Martha Nussbaum argues that literature can increase empathic understanding in a way that serves democracy, namely by increasing our powers of imagination in representing all kinds of people with circumstances and problems different than our own.[25] Drawing on Sophocles, Philoctetes, and Ralph Ellison, Nussbaum contends that insofar as literature involves us in the inner worlds of its characters, we learn from it how to see things from another perspective.[26] Literary imagination, then, is crucial for cultivating sympathy and understanding toward invisible and marginalized people and offers a view toward a civic imagination. Literature is seen to make us better people, because it renders us more sympathetic to others and more sensitive to the frailties of our own moral intentions and actions. The Enlightenment hope for an aesthetics that would serve the higher goals of humankind survives in such notions that literature could make us more moral and more properly human.

All of these views hinge on tragedy's effects upon an audience. They portray the sequence of tragic events in terms of its pedagogical value in evoking audience insight and feeling. What if instead catharsis can be seen as pertaining not to the emotions of the audience but to the events themselves in their connection? In O.B. Hardison's "deflationary" interpretation and Leon Golden's translation, catharsis is not understood as a matter of emotional cleansing, de-polluting, sentimental education, nor even our recognition of our kinship with the hero in his finitude. In Golden's translation, catharsis instead names the achievement of a *clarification* (not a purgation) and what is clarified is the course of *connected incidents* that forms the plot of the tragedy.[27] Hardison proposes that the *catharsis* of τῶν τοιούτων παθημάτων refers not to emotions at all, but instead to such a *clarification* of emotion-filled *incidents*. In the Golden/Hardison reading

---

Aristotelian theory of tragedy is seen as a tragedy of fate, in contrast to the later pre-Romantic tragedies of character (436). But *Philotas* and *Emilia Galotti* perform, in Brodsky's careful readings, undoings of the Aristotelian prescriptions for tragedy (442–446).

24 Richard Rorty, *Contingency, Irony, and Solidarity* (New York: Cambridge University Press, 1989), 141.

25 Martha C. Nussbaum, *Cultivating Humanity: A Classical Defense of Reform in Liberal Education* (Cambridge: Harvard University Press, 1998), 95–96; 86.

26 Nussbaum, *Cultivating Humanity*, 87.

27 Hardison, "Commentary," 116.

catharsis involves only the disclosure of an inexorable connectedness of the events of the tragedy. Rather than relating to the psychology of the spectator and the purging of emotions, the catharsis *of tragic incidents* entails an enhanced depiction of a necessary and probable course of events. It demands, in other words, the revelation within the incidents themselves of their maximal connection to each other.[28]

M.D. Petrusevski offered an even more extreme version of this hypothesis, on the basis of a philological claim. As Claudio Veloso and others point out, Petrusevski argues that the crucial Greek phrase *pathematon katharsis* – usually understood as the catharsis of feelings, but for Golden and Hardison read as the catharsis of emotion-filled incidents – is not properly part of Aristotle's text *at all*. Instead, the phrase in question should read *pragmaton sustasin*, "the arrangement of incidents."[29] Petrusevski and others point out that the phrase in question makes much more sense when read as a reiteration of the text's own declaration that the arrangement of events is the point of tragedy, rather than as an introduction of a "purgation" theme that is otherwise peripheral to the *Poetics*.[30] Hardison similarly points out that the interpretation of catharsis as related to emotion-filled *events* rather than to the emotions themselves has the advantage of consistency with Aristotle's emphasis throughout the *Poetics* on the primacy of plot. Golden and Hardison draw attention to Aristotle's mention at 1449b23 that the definition of tragedy, in which the catharsis passage appears, draws on "what we have already said."[31] They note that the preceding chapters of the *Poetics* say nothing whatsoever about catharsis, pity, or fear. For Golden and Hardison, and in a different way for Petrusevski, the purgation of emotions on the part of the audience simply does not fit with the rest of the *Poetics*. Martha Husain argues, in line with Golden and Hardison, "The *katharsis* to be achieved is the

---

28 Lear, "Katharsis," 1988, 339n79, contends that Golden is wrong, but does grant that there could be both a subjective and objective catharsis, one related to emotions and one related to a ritual sacrifice in the play.
29 We retain the English transliteration here in echo of Petrusevski's usage. M.D. Petrusevski, "Pathematon Katharsin ou bien Pragmaton Systasin?" *Ziva Antika/Antiquité Vivante* 4, no. 2 (1954): 209–250; Petrusevski, "La définition de la tragédie chez Aristote et la catharsis," *L'Annuaire de la Faculté de philosophie de l'Université de Skopje* 1 (1948): 3–17. See also Gregory Scott, "Purging the *Poetics*," in *Oxford Studies in Ancient Philosophy* 25 (2003): 234; Svetlana Boym, *Another Freedom: The Alternative History of an Idea* (Chicago: University of Chicago Press, 2010), 308n23.
30 Claudio William Veloso, "Aristotle's *Poetics* without Katharsis, Fear or Pity," *Oxford Studies in Ancient Philosophy* 33 (2007): 268.
31 Hardison, "Commentary," 114.

clarification of the action's sequential-causal structure" and hence "[t]he definitional achievement of tragedy lies not in denying the emotive content of the action but in making it causally comprehensible."[32] Alexander Nehamas proposes to interpret *catharsis* as the "solution of the tragic plot," "resolution," or "dénouement," rather than as a purgation.[33] Nehamas emphasizes the punctual ending or closure of the tragedy as the site of clarification.

If catharsis is at the center of Aristotle's definition of tragedy, then in the readings of Golden and Hardison, Veloso and Petrusevski, Husain, and Nehamas, that center is not primarily an emotional one. It instead is a matter of how a course of events unfolds according to their inexorable connections. This does not negate the importance of emotion entirely. The incidents themselves, to be sure, are fearsome and pitiable. Nonetheless, the emphasis in the above-mentioned readings on the inexorably unfolding events downplays the importance of subjectivity – both of the hero and of the audience. What is at stake instead is a terrifying automatism – i.e., the relentless unfolding of a sequence of events beyond human control or intention.

Suspense for the sake of suspense

The interpretations described above that focus on connected events do not exclude the possibility that tragedy is more philosophical than history because it provides greater insight into our human finitude in the face of inexorability. Here, Golden and Hardison allow that tragedy bears a pedagogical value: such portrayals of inexorable connections may instruct us about our helplessness in the face of fate or causality.[34] The clarification of the connections of the tragic events thus produces what they call an "insight experience."[35] We *experience* in tragedy the inexorability of events that are intrinsically connected and hence achieve *insight* into our own finitude.

---

32 Martha Husain, *Ontology and the Art of Tragedy: An Approach to Aristotle's Poetics* (Albany: SUNY Press, 2012), 45.
33 Alexander Nehamas, "Pity and Fear in the *Rhetoric* and *Poetics*," in Rorty, *Essays on Aristotle's Poetics*, 307.
34 Nussbaum, *Fragility of Goodness: Luck and Ethics in Greek Tragedy and Philosophy* (Cambridge: Cambridge University Press, 2013), rejects the notion of intellectual clarification and suggests that instead clarification occurs through our emotional response. The emotional response "leads us to understand what our values are" (390).
35 Hardison, "Commentary," 117.

G.W.F. Ferrari argues, however, that insight, empathy, and edification are entirely beside the point. Instead "[t]ragedy is plot for plot's sake,"[36] and Aristotle's *Poetics* constitutes a recipe for sheer pleasure and maximum entertainment. For Aristotle, according to Ferrari, tragedy is purely a matter of the art of suspense.[37] This suspense depends solely upon the artful exposition of the connections between the incidents. Following Dorothy Sayers's treatment of Aristotle and detective fiction, Ferrari rejects all didactic and moralizing interpretations of the *Poetics* and proposes that for Aristotle, tragedy has nothing whatsoever to do with empathy or finitude.[38] Ferrari's reading is important to the present argument because its insistent fixation on the emplotted connection between events serves drastically to reformulate the significance of the *Poetics* as a cornerstone in the history of literary aesthetics. Empathy, insight, profundity, and even philosophy per se are not intrinsic to Aristotle's purpose. Instead, "all that Aristotle is [...] doing [...] [is analyzing] the lowest common denominator of the pleasure we take in fictions."[39] Suspense per se is the highest value, in this reading, because it produces maximal pleasure for an audience, hence the representation of unexpected connections is what is central.

As do Golden and Hardison, Ferrari points to Aristotle's stress on the construction of plot as the primary task of the poet – albeit without any Enlightenment-style allowance for the significance of insight into our own vulnerability. Ferrari shares with Armstrong the notion that the "universality" of tragedy is not a matter of truth nor of insight but refers instead to the discerning construction of the plot. Hence

> the universality at which poetry aims is not its ultimate goal but something it must look to if its plots are to be properly constructed. [...] Poetry "says what is universal" not by making universal statements but by fulfilling the task that Aristotle assigned to the poet [...] [i.e.,] saying "what could be expected to happen."[40]

"Universality" here is seen as having no moral or didactic connotation. It is instead a matter of the sheer likelihood that the events would happen as they do in the tragedy, in accord with our expectations. As with Armstrong, it is not the

---

36 G.W.F. Ferrari, "Aristotle's Literary Aesthetics," *Phronesis* 64, no. 3 (1999): 183. For Ferrari, Aristotle is thoroughly concerned with "the qualities of what we would now call 'literature' and [with] investigat[ing] what makes a literary work of art successful in its own special terms – with tragic drama, the consummate literary form, selected for particular attention."
37 Ferrari, "Aristotle's Literary Aesthetics," 183.
38 See Dorothy L. Sayers, "Aristotle on Detective Fiction," *English* 1, no. 1 (1936): 23–35.
39 Ferrari, "Aristotle's Literary Aesthetics," 185.
40 Ferrari, "Aristotle's Literary Aesthetics," 184

tragedy nor its effect on the audience that Ferrari finds philosophical; likewise "universal" has nothing to do with an effect, truth, or teaching that might be ascribed to tragedy. Rather, universality refers to a criterion for proper plot construction. Ferrari interprets Aristotle's emphasis on universals as a reference to the importance of constructing a plot according to what could be *expected* to happen. The construction of a plot requires an understanding of the universal, and hence philosophical, principles that connect events. While other critics puzzle over Aristotle's loose use of key terms related to τὸ ἀναγκαῖον in the description of how the events should be connected, arguing over whether they refer to probability, necessity, likelihood, or plausibility, for Ferrari the selection of the precise term is beside the point. The point is that the tragic events are connected, "and never mind exactly how."[41] Connection per se is the heart and soul of tragedy; and therefore, it misses Aristotle's point to impose fine-grained philosophical analyses of types of connection, universal truths, and empathy, for maximal suspense alone is that with which the *Poetics* is concerned.

In his singleminded focus on suspense, Ferrari dismisses the preoccupation of "didactic" interpretations of Aristotle's references to mimesis, pleasure, and learning [μανθάνειν] in the *Poetics*' early sections. With regard to "learning things gives great pleasure" (1448b16–17) Ferrari translates μανθάνειν as "figuring out" and casts the context as a matter of imagination, as "what happens when the audience accepts the fiction-maker's invitation to imagine a state of affairs."[42] Likewise illuminating is Ferrari's take on Aristotle's reference to the pleasure we take in seeing imitations of corpses or frightening animals (1448b12). Aristotle's reference to the corpse has been understood canonically as making a significant connection between pleasure and learning:

> [I]t is an instinct of human beings, from childhood, to engage in mimesis (indeed [...] man is the most mimetic of all, and it is through mimesis that he develops his earliest understanding); and equally natural that everyone enjoys mimetic objects. [...] [W]e enjoy contemplating the most precise images of things whose actual sight is painful to us, such as the forms of the vilest animals and of corpses. The explanation of this too is that understanding gives great pleasure. (1448b5–12)

This passage has lent itself to interpretations that Ferrari finds overly philosophical, e.g., propositions that Aristotle is reflecting on the existential value of mimesis or on our relationship to death, or that he ties the pleasure in tragedy to the pleasure in seeing simulations of corpses. For Ferrari, these renderings

---

41 Ferrari, "Aristotle's Literary Aesthetics," 189.
42 Ferrari, "Aristotle's Literary Aesthetics," 186.

overread the *Poetics*. The mention of the corpse in this early section of the book does not prefigure or otherwise pertain to the deaths that take place in tragedy. Tragedy is, in turn, not a more complex version of the pleasure of viewing frightening or disgusting objects or animals. Rather, Aristotle takes these extreme examples of viewing simulated corpses and vile animals to prove, precisely because they would seem the *least* likely to evoke pleasure, that we take pleasure in acts of imagination per se. The corpse and the frightful animal are striking and illuminating instances of how, even where the content is objectionable, we derive pleasure from the imaginative process of viewing it: "It is an extreme case to show that the basic pleasure of mimesis comes from the act of imagination rather than depending on the nature of the object imagined."[43] In tragedy, therefore, the suspense-releasing disclosure of connection makes for greater pleasure because viewing it is a greater accomplishment of imagination than viewing a static image. It has nothing to do with learning nor with the recognition of finitude, it is about the pleasure of a suspenseful climax.

Ferrari resists at each turn a didactic, moral interpretation of the *Poetics*. Aristotle is instead seen as providing recipes for suspenseful – and only for that reason good – literature. Hence for Ferrari, the *Poetics* is sheerly a matter of literary aesthetics, of excellent design, with no didacticism and no truth on offer. The connection of events and the pleasure taken therein is the only point and the events' suspense-inducing arrangement is the highest achievement of the tragedian. For this reason, Ferrari treats ἀναγκαῖον – understood variously, as has been said, as necessity, likelihood, or probability in the connection of events – as an aesthetic rather than a moral term.[44] Tragedy is likewise not about fate's influence on human affairs. Rather, the writer designs a plot that produces maximal suspense and release by means of its coherence, excellent arrangement, and unity. The relentless inevitability of the succession of events in the tragedy is "a fictional creation of the dramatist's," and the design and artistry are its crucible.[45] In this radically aesthetic interpretation of the *Poetics*, the pleasure of tragedy has nothing to do with insight. Rather, it involves amazement or marvelousness, which is a matter of "the intricacies of the plot, and the winding-up and subsequent release of suspense."[46] In this interpretation, Aristotle provides no support for claims that literature makes us wiser, better, or more moral. Moreover, tragedy does not serve as a testing ground for existential questions of morality and literature per se. Such readings that tend toward existential

---

43 Ferrari, "Aristotle's Literary Aesthetics," 186
44 Ferrari, "Aristotle's Literary Aesthetics," 187.
45 Ferrari, "Aristotle's Literary Aesthetics," 188.
46 Ferrari, "Aristotle's Literary Aesthetics," 186.

implications put "more moral and existential weight on the girders of Aristotle's structure than they were designed to bear."[47] The unexpected connectedness of events is not meant to teach the audience anything at all.

As evidence that interpreters have imported existential and moral concerns into a sheerly aesthetic treatise, Ferrari points to the dominant interpretations of *Poetics* 1452a4, in which Aristotle first explains the fear- and pity-inducing connection of events and then ties this to a sense of awe and wonder at events that seem to happen according to an overall plan:

> Given that the mimesis is not only of a complete action but also of fearful and pitiable matters, the latter arise above all when events occur *contrary to expectation yet on account of one another* [ταῦτα δὲ γίνεται καὶ μάλιστα [καὶ μᾶλλον] ὅταν γένηται παρὰ τὴν δόξαν δι' ἄλληλα]. (1452a4, emphasis mine)

Canonical literary theory emphasizes as the key to tragedy's pleasure the conjunction of unexpected events on the one hand and causal connection on the other hand. Most translations include in the key clause an adversative – "yet" or "but" – in order to indicate that there is a constitutive conflict in tragedy between the causal connection of the events (because of one another) and the unexpected element (contrary to expectation). The adversative emphasizes that two distinct elements bump against each other, i.e., the events' unexpectedness and their causal connection. Hence Halliwell's translation, as we see above, suggests that the tragic incidents "occur contrary to expectation *yet* on account of one another," Golden translates the identical portion as "occur unexpectedly, *yet* because of one another," and Richard Janko translates it as "contrary to expectation and *yet* because of one another."[48] In each of these translations, the "yet" emphasizes that on the one hand the tragic *events* are what is described as unexpected and *in addition* they are portrayed as causally connected.

Ferrari, however, points out that Aristotle's Greek contains no adversative at all, no "yet," "but," or "still."[49] Against all of the interpretations above, he understands Aristotle to be saying that the events are not *in themselves* unexpected or astonishing, but rather *their connection to one another* is what is astonishing and unexpected. He translates the passage as one phrase with no adversative, i.e., as "contrary to expectation because of one another." In this

---

47 Ferrari, "Aristotle's Literary Aesthetics," 190.
48 Aristotle, *Poetics* [Halliwell translation], 63; Leon Golden, *Aristotle's Poetics: A Translation and Commentary for Students of Literature* (Tallahassee: Florida State University Press, 1981), 18; Richard Janko, *Poetics, with the "Tractatus Coislinianus," Reconstruction of Poetics II, and the fragments of the "On Poets"* (Indianapolis: Hackett, 1987), 13. Emphasis mine in all cases.
49 Ferrari, "Aristotle's Literary Aesthetics," 190–191.

interpretation, the direct causal *connection* between the events is what is unexpected, not the events themselves. The suspense and release of tragedy are achieved with the revelation of the *unexpected causal connection* between the events, rather than with the occurrence of unexpected *events* that are *also* causally connected. What is significant in each case is that "the pieces unexpectedly come together," according to Ferrari.[50] That is, events in a good plot "defeat expectation *by virtue of* resulting from one another," rather than owing to a combination of the unexpectedness of the events and their causal connection.[51] That things unexpectedly fit together in a tragedy is the source of its great pleasure.

This explains why, for Ferrari, there is an act of imagination, rather than learning, that produces the pleasure of tragedy. In his interpretation, the pleasure of tragedy derives from the imaginative act that resolves the tension between the actual connectedness and the seeming disconnectedness of the series of events that unfolds in the tragedy. It is significant for Ferrari that Aristotle points out immediately following "contrary to expectation because of one another" that even events that *merely appear* connected by design produce a pleasurable astonishment. The Mitys anecdote, in which Mitys's killer was himself killed when a statue of Mitys fell on him, is instructive:

> The awesome will be maintained in this way [i.e., when events occur contrary to expectation because of one another] more than through show of chance and fortune, because even among chance events we find most awesome those which seem to have happened by design (as when Mitys' statue at Argos killed the murderer of Mitys, by falling on him [...] such things seem not to occur randomly). And so, such plots are bound to be finer. (1452a4–10)

Unexpected connection, rather than randomness, chance, and fortune, produces a particular pleasure. For Ferrari, Aristotle's treatment of the Mitys anecdote shows that tragedy has nothing to do with learning anything about fate or human finitude. Instead, the anecdote's appearance in the *Poetics* confirms that even an unexpected *appearance* of connection produces a great pleasure in both tragedy and in life. For it *seems* as though the statue's fall onto Mitys's murderer *must* somehow represent an act of Mitys's revenge, or at least that it must represent "poetic justice." The coincidental sequence of events looks *as though* it had happened according to a plan or plot – as though the statue's fall were *connected* to the crushed man's act of murder.[52] The point is that

---

50 Ferrari, "Aristotle's Literary Aesthetics," 192.
51 Ferrari, "Aristotle's Literary Aesthetics," 191, *sic* italics.
52 The story of Mitys's killer "seems [...] too good to be true; seems, as we might also say, just like a story." The killer's death-by-sculpture appears meaningful even though it forms no part of a story plot (Ferrari, "Aristotle's Literary Aesthetics," 192).

surprising connections, and *even mere appearances* of surprising connection (also known as coincidences), are striking – to the characters, to audience members unacquainted with the story, and to those who use "that capacity we all have to find a novel thrilling even on second, third, fourth readings."[53] The tension and release of observing the buildup of events, along with the discovery *of their connection*, is the source of the pleasure and the goal of tragedy. Aristotle writes that tragedy needs to produce a sense of awe or marvel in order to be entertaining (1460a12). When events occur because of one another and yet *this relationship* is unexpected, this is for Aristotle "more marvelous" than sheer chance.[54] This is not about philosophical insights on the part of the audience. It is the writer's hand that crafts the coincidence that evokes the tragic pleasure: "for every drama and literary fiction tout court [...] [t]he human dramatist plays god in the little world he has created."[55]

Again, Ferrari's anti-existential reading rejects the idea that tragedy should illuminate our own existence within finitude, chance, and inexorability. Rather, it is a matter of the pleasure we get from the unexpected connections of the tragedy – a pleasure not of insight (despite Aristotle's references to imitations, learning, and inferring) – but of suspense and release. Fear and pity are therefore not existentially more important than other emotions. For instance, tragedy might arouse anger and shame as well.[56] Fear and pity are nonetheless crucial simply because they happen to be, in Aristotle's view, the ingredients for maximal suspense and release in the audience. What is more, if Aristotle is making purely literary prescriptions for properly tragic pleasure, with no existential, intellectual, ethical, or edifying implications, then the entire tradition of literary criticism has overread the significance of the *Poetics*.

The riven *Poetics*

Insofar as Aristotle requires that tragedy contain a "single, whole, and complete action" (1459a16), unity and wholeness are important themes in the *Poetics*.[57]

---

53 Ferrari, "Aristotle's Literary Aesthetics," 192.
54 The only "larger" take on tragedy is that tragic plot (as in *Oedipus Rex*) "effects a match between the machinations of the gods and those of the human mechanic who makes the whole thing run – the playwright" (Ferrari, "Aristotle's Literary Aesthetics," 193).
55 Ferrari, "Aristotle's Literary Aesthetics," 193.
56 Ferrari, "Aristotle's Literary Aesthetics," 194.
57 Obviously, the issue of unity and completeness is prominent in the late eighteenth-century European development of aesthetics as a distinct area of inquiry, in which the sublime as

The question of what constitutes a whole is central in the history of modern German aesthetics and its origins in early modern metaphysics, which is in turn rooted in Aristotelianism.[58] Leibniz's theological metaphysics, for instance, provides a useful contrast to the Aristotelian significance of connection for tragic unity. For Leibniz, each substance unfolds itself in accord with its own essence. What we perceive as interactions *between* substances, including the apparent connections of cause and effect, are predetermined by God's decision to create the universe and each substance in the way that he did. For this reason there is no tragedy as such in the larger scale of Leibnizian metaphysics. Finite human beings see only a portion of the universe, but God maintains the overview, unavailable to the human faculties, of its goodness and unity.[59] In the production of tragedy, in contrast, unity and wholeness depend on the thoroughgoing connections of events in a one-by-one fashion and according to the rules for tragic plot.

What about the unity of the *Poetics* itself? Throughout the reception of the *Poetics*, the question of its continuity and coherence has been decisive. Its interpreters attempt to account for disparate elements of the text – the Mitys anecdote, the references to corpses and vile animals, the emphasis on universals – in ways that confirm their particular interpretations of the *Poetics* as a whole. As we have just seen in the previous section, there exist drastically different explanations of the connections between the *Poetics*' opening discussion of mimesis in

---

described by Edmund Burke and Kant involves the attempt to conceive of a whole that is too large to take in at once. In a fascinating prefiguration of this, Aristotle makes the seemingly odd claim that a very small animal cannot be beautiful; the implication is that the reason for this is that it is impossible with something so small to see how the parts relate to the whole, while nonetheless because it is an animal it is by definition a whole and unitary entity. Aristotle further indicates that an animal so huge that we could not see all of it would not be beautiful because we could not see it as a unity (1450b33–1451a2).

58 Aristotle's prescriptions for the unity of tragedy describe in some detail what it means to portray a complete action, namely that it must have a beginning, middle, and end. Armstrong points out that in the *Physics*, Aristotle describes the continuity of processes in similar terms: "[A] process (*kinesis*) is one if and only if it is continuous." Similarly, Armstrong points out that for Aristotle's *Physics*, "two events are continuous if and only if their 'extremities' are one (228a20–4; 227a10–12). [...] They [the extremities] must go together out of necessity (227a23)." Armstrong, however, asks whether necessity is in conflict with the probability of connected events in tragic plot. He concludes that the necessity that governs the *Physics* is less stringent than the probability that governs the *Poetics*, but they are allied in that in both cases the opposition is to chance (Armstrong, "Philosophical Nature of Poetry," 453–454).

59 See Hans Blumenberg, "The Concept of Reality and the Possibility of the Novel," in *New Perspectives in German Literary Criticism: A Collection of Essays*, eds. Richard E. Amacher and Victor Lange, trans. David Henry Wilson (Princeton: Princeton University Press, 1979), 29–48.

its early sections and the prescriptions for tragedy that make up the bulk of the text. David Ferris's radical reinterpretation of the *Poetics* takes the issue of continuity in another direction, one that also indicts the whole of literary criticism, but on completely different grounds than Ferrari's critique of existential and philosophical readings. Ferris argues that the entire history of reception of the *Poetics* recapitulates an evasion that is performed in the text itself. As does Ferrari, Ferris rejects traditional views that the discussion of mimesis and learning are intrinsically connected to the prescriptions for tragedy. Unlike Ferrari, however, who sees the discussion of mimesis as a minor issue, Ferris argues that there is a profound and meaningful *dis*connection between the discussion of mimesis and the description of tragedy and its components. That disconnection indicates that the *Poetics* itself does not form a proper or cohesive unity. Ferris points out that in its first five chapters, the *Poetics* raises the question of mimesis and attempts to define it with reference to various art forms and discussions of melody, meter, rhythm, and dance. Aristotle nonetheless fails to achieve a definition of mimesis. The topic of the treatise then swerves at 1449b21 from the question of mimesis to the definition of tragedy as if the explanation of mimesis had been attained.[60] In fact, however, Ferris finds that the text "has been unable to provide an adequate account of its central concept" of mimesis.[61] The *Poetics* thus evades the problem of mimesis *and* represses its own failure to answer the question of precisely what mimesis is. The entire treatment of tragedy, Ferris argues, constitutes an evasion of that question, rendering the text itself profoundly riven.

Ferris's argument also expands to encompass the entire canon of literary theory. He proposes that the *Poetics'* opening sections on mimesis, and in particular their disconnection from the much longer discussion of tragedy that follows, bear an unrecognized significance for both the *Poetics* and its legacy. The treatment of mimesis and the description of tragedy are canonically interpreted as being commensurate and mutually illuminating. The history of literary theory, Ferris claims, suppresses the *Poetics'* critical *dis*junction between the opening chapters on mimesis and the definition of tragedy. Literary theory passes over the breach in the *Poetics* in order to produce its own coherent reception of literature. Literary theory, in other words, treats the cornerstone text *as if* it were an intact unity. The rivenness of the text is not overcome, however, and

---

60 David Ferris, *Theory and the Evasion of History* (Baltimore: The Johns Hopkins University Press, 1993), 28.
61 Ferris, *Evasion of History*, 26.

the unanswered question of mimesis haunts the entire subsequent reception of the *Poetics* and the canon of literary theory.

In a similar argument, Samuel Weber borrows from Freud's understanding of how dreams are recounted after waking in order to understand the riven character of the *Poetics* and its legacy in literary theory. According to Freud, a coherent account of a dream produced by a patient conceals less coherent and more disruptive unconscious operations.[62] While the ego creates daydreams that have coherent narratives, dreams in sleep lack this narrativizing function.[63] Only in secondary revision, once the patient is awake, does the dream achieve such wholeness. Weber argues that Freud's description of a logic of secondary revision applies in numerous ways to the reception of the *Poetics*. The text itself is "reconstituted long after the fact, apparently from notes, and [is] anything but simply complete or finished."[64] The legacy of literary theory nonetheless treats the *Poetics* as a complete and consistent text:

> [D]espite its systematic-sounding claims, [the *Poetics*] is in fact a fragment and by no means an entirely coherent or completed work. [...] [S]uch incoherence is often ignored, as though to take notice of it would be to reduce the text's value.[65]

The reception of the *Poetics*, like the dreamer recounting a dream upon waking, retroactively reconstitutes the inconsistent and fragmentary text as an intelligible whole. Weber further argues that several sorts of secondary revision occur in Aristotle's prescriptions for plot. Specifically, the *Poetics'* thematization of the unity of plot covers over or compensates for the thoroughgoing dividedness of the theatrical medium itself. For instance, Weber points to the divisions between a particular iteration of events that takes place on the stage and the constancy of the work itself, between the actors onstage and the characters they represent, between the *opsis* or spectacular means of theater and the *synopsis* or overview it should present, and between theatrical and narrative representation.[66] The medium of theater and the narrative of tragedy are represented in the *Poetics* in ways that obscure both their heterogeneous elements and the profoundly open-ended character of the theatrical medium.

For Weber, in other words, the "distinguishing features" of theater are eclipsed by Aristotle's emphasis on plot in the *Poetics*. The specific mediality of

---

62 Samuel Weber, "Psychoanalysis and Theatricality," *Parallax* 6, no. 3 (2000): 39.
63 Weber, "Psychoanalysis and Theatricality," 41.
64 Samuel Weber, *Theatricality as Medium* (New York: Fordham University Press, 2004), 99.
65 Weber, "Psychoanalysis and Theatricality," 32.
66 Weber, "Psychoanalysis and Theatricality," 35.

theater and the divisions it harbors are subordinated to the genre of poetry that depends on narrative.[67] Hence

> the significance of events, persons, and things, depends not on their inner, self-identical substance, but on their situation, literally, on their placement, on their relation to others. And [...] such situatedness, such placement, can never be *finished*, in the sense of being completed, but only *interrupted* and redeployed.[68]

The *Poetics*' prescriptions for plot and the elements of tragedy demand completion but theater itself is by nature *in*complete. Both the text itself and its receptions pass over its many points of incompleteness when it comes to laying out the prescriptions for the medium of theater. Analogously to Ferris's argument about the abandonment of the central concept of mimesis, Weber suggests that in its preoccupation with plot the *Poetics* covers over its own failure to complete the task with which it began, namely of defining the operations of theater.

The issues of continuity and connection between individual elements that we have observed in various readings of the *Poetics* can, in light of these readings by Ferris and Weber, be raised regarding the *Poetics* itself. Perhaps the text is held together in and by the legacy that elides the profound discontinuities and incongruities that it contains. The conventional focus in the *Poetics*' reception on insight, feeling, and universality eclipses the text's own insistence on the connections between events. Those connections in turn stand in tension with the unitary structure that should constrain them. The heterodox views of Armstrong, Golden and Hardison, and Ferrari on universality, catharsis, and suspense speak against a unitary legacy of the *Poetics*. Ferris's argument about the breach within the text and its evasion in literary theory accentuates that contradiction. Likewise Weber's insight about the significance of the theatrical situation of tragedy can be applied to the *Poetics* itself: the text's significance does not depend on a unitary message or "inner, self-identical substance."[69] Rather its significance arises in its relation to other concepts over the history of its reception – the concepts of catharsis, universals, and plot. The *Poetics* itself, in other words, bears an open-ended, unfinished significance, one that is always interrupted and repurposed in further readings. Its receptions are themselves secondary revisions, creating a unity where there is no unified original. This is ironic insofar as within the *Poetics*, as we have seen, Aristotle attempts to describe what makes for a unified whole. Its receptions render the *Poetics*

---

67 Weber, "Psychoanalysis and Theatricality," 39.
68 Weber, "Psychoanalysis and Theatricality," 38.
69 Weber, "Psychoanalysis and Theatricality," 38.

a unified whole based on certain values they search for within it – for instance universal truth, insight, and our common humanity. In making Aristotle into a unitary origin, and making itself the recipient and successor of an orthodox legacy, literary theory renders the *Poetics* a unitary whole despite the wildly heterogeneous receptions that inhabit and dispute the canon of literary theory.

2 Contingency, Connection, and Possible Worlds: History and Poetry in Gottsched's *Versuch einer critischen Dichtkunst*

Johann Christoph Gottsched serves as a poster child of the eighteenth-century German legacy of Aristotle's *Poetics*, in particular in his attempts to catalogue rules for poetry.[1] Indeed, Gottsched has long been read, and dismissed, within German Studies as a stuffy prescriber of Aristotelian poetic rules and recipe-driven composition, thanks in part to Lessing's mocking attacks in his seventeenth *Literaturbrief*.[2] The emphasis on reason in his formulation of poetic prescriptions has also produced a view of Gottsched as, in Frederick C. Beiser's terms, "the high noon of aesthetic rationalism."[3] If we see Gottsched in the exclusive light of the rationalist philosophical tradition, however, we pass over certain central elements of Gottsched that do not fit this characterization. Such a rationalist caricature obscures the extent to which his account of poetic connection ultimately portrays the associations that make up poetic thinking as wildly ungovernable and heterogeneous.

The caricature of Gottsched as "the high noon of aesthetic rationalism" and as a mouthpiece of Aristotelian prescriptions also obscures the significant ways

---

1 See Andreas Härter, *Digressionen: Studien zum Verhältnis von Ordnung und Abweichung in Rhetorik und Poetik: Quintilian, Opitz, Gottsched, Friedrich Schlegel* (Paderborn: Fink, 2000); Gerhard Schäfer, *"Wohlklingende Schrift" und "rührende Bilder": Soziologische Studien zur Aesthetik Gottscheds und der Schweizer* (Frankfurt am Main: Lang, 1987); Rosemary Scholl, "Die Rhetorik der Vernunft: Gottsched und die Rhetorik im frühen 18. Jahrhundert," *Jahrbuch für Internationale Germanistik* 2, no. 3 (1976): 217–221; Katherine R. Goodman, "Gottsched's Literary Reforms: The Beginning of Modern German Literature," in *German Literature of the Eighteenth Century: The Enlightenment and Sensibility*, ed. Barbara Becker-Cantarino (Rochester, New York: Camden House, 2005), 55–76; Hans Freier, *Kritische Poetik: Legitimation und Kritik der Poesie in Gottscheds Dichtkunst* (Stuttgart: Metzler, 1973); Uwe Möller, *Rhetorische Überlieferung und Dichtungstheorie im frühen 18. Jahrhundert* (Paderborn: Fink, 1983). For a thorough treatment of the role of mimesis and its relevance to a broader view of Gottsched, see Sarah Ruth Lorenz, "Shifting Forms of Mimesis in Johann Christoph Gottsched's *Dichtkunst*," *German Quarterly* 87, no. 1 (2014): 86–107.
2 Gotthold Ephraim Lessing, "Briefe, die neueste Literatur betreffend," in *Werke*, eds. Herbert Göpfert et al. (Munich: Hanser, 1970–1989), 5:70–73. See also Brigitte Peters, "Der 17. Literaturbrief und seine Folgen," *Zeitschrift für Germanistik* 10, no. 1 (1989): 70–75.
3 Frederick C. Beiser, *Diotima's Children: German Aesthetic Rationalism from Leibniz to Lessing* (Oxford: Oxford University Press, 2009), 72. Beiser argues that Kant's stress on the non-cognitive aspect of aesthetic judgments came to overshadow, regrettably and for the entire subsequent reception of aesthetics, the pre-Kantian concern for objective principles of taste (16).

in which Gottsched blurs the boundaries between poetic writing and historical writing, even as he attempts to distinguish them from one another. That is, Gottsched articulates the distinctions between history and poetry according to Aristotelian prescriptions, but defines differently than Aristotle the principles by which events are connected to one another in each type of narrative. This chapter will argue that Gottsched's poetic prescriptions can be viewed less as rational rules than as responses to the problem of managing a proliferation of connections that result from a labile, even illimitable, poetic possibility. For poetic thinking, according to Gottsched, does not operate according to logical rules, but instead involves a promiscuous connectibility of thoughts that actually subverts rule-giving as such, above all by way of a prolific acuity [*Scharfsinnigkeit*].[4] The rational rules that Gottsched formulates ostensibly should serve to govern the connection of thoughts in poetry, but as he describes how thoughts connect to one another, the rules actually attest to the *ungovernability* of poetic thinking.[5]

In particular, as we will see, Gottsched describes poetic thinking as rooted in an indiscriminate connectibility that associates thoughts by way of memory, similarity, feeling, and, most indiscriminate, *possible* similarity. The ungovernable connection of thoughts along such elastic lines is represented as a matter of figurality, born out of the poet's *Scharfsinnigkeit*, which renders the connections between thoughts less a matter of a temporal process than of near-instantaneous constellation. Nonetheless, the labile *Scharfsinnigkeit* that animates poetic thinking must be transformed into a poetic style of writing [*poetische Schreibart*], which for Gottsched means that the poetic elements must clearly fit together and make sense in their connections to one another. Here the term *Fabel* is key. Commenting on Aristotle's concept of μῦθος, Gottsched pronounces *Fabel* the essence of poetry, as what connects the elements, "*die Zusammensetzung*

---

4 *Scharfsinnigkeit* has also been translated as "astuteness," see Eric Williams, *The Mirror and the Word: Modernism, Literary Theory, and George Trakl* (Lincoln: University of Nebraska Press, 1993), 42; and Matthew Bell, *The German Tradition of Psychology in Literature and Thought, 1700–1840* (Cambridge: Cambridge University Press, 2005), 30.
5 As Benjamin Bennett, *Beyond Theory: Eighteenth-Century German Literature and the Poetics of Irony* (Ithaca: Cornell University Press, 1993), 232, points out, for Gottsched (and Bodmer and Breitinger) the formulation of rules did not have as its goal, as in the French tradition of poetic rules, to offer rules of application to a fixed canonical language. Instead, for the early eighteenth-century German treatment of poetic language, the question was, Bennett writes, what it means to have a common language in the first place. This implies that Gottsched's attempts to formulate rules must be understood in the context of the development of German as a literary language.

oder Verbindung der Sachen [the composition or connection of the matters]."⁶ *Zusammensetzung* is not happenstance, but must reflect an inherent, integral connection. Poetry must, in other words, follow a principle of real connection, or *connexio realis*. The reference to *connexio realis* in the context of poetry speaks against the Aristotelian distinction, which Gottsched nonetheless echoes, between the reality that defines the content of historical narrative and the possibility that defines the content of poetic narrative. As we will show, Gottsched's Wolffian characterization of *Fabel* as a story from another world confounds an absolute delimitation of possibility and reality. Gottsched's attempt to keep history and poetry apart, in line with Aristotle's dictum, falters in other ways as well. This is perhaps most obvious where Gottsched dictates that the history-writer [*Geschichtschreiber {sic}*] must *borrow* poetic techniques. The ambiguity in Gottsched's understanding of borrowing exhibit what Reinhart Koselleck in another context has characterized as "the demands of history and poetics folded together."⁷ Gottsched is on the cusp of their *un*folding, but his *Dichtkunst* inscenates a tangle of possibility, reality, history, and poetry.

Borrowing from the possible

As we have seen in Chapter One, in the *Poetics* Aristotle makes a stark distinction between poetry and history, on the basis of the possibility or reality of their respective subject matters:

> [I]t is not the poet's function to relate actual events, but the *kinds* of things that might occur and are possible in terms of probability or necessity. The difference between the historian and the poet is not that between using verse or prose. [...] [T]he difference is this: that the one relates actual events, the other the kinds of things that might occur. (1451a35–39)

This passage is the source of a canonical distinction that is presumed to govern pre-nineteenth-century writing. It provides the background, for instance, to Hayden White's claim that until the nineteenth century, the distinction between historical and literary writing was unproblematic.⁸ Rüdiger Campe, however, has argued for the development, as early as the seventeenth century, of a gradation

---

6 Johann Christoph Gottsched, *Versuch einer critischen Dichtkunst*, vol. 6, bk. 1, *Ausgewählte Werke*, eds. Joachim Birke and Brigitte Birke (Berlin: De Gruyter, 1973), 203.
7 Reinhart Koselleck, *Futures Past: On the Semantics of Historical Time*, trans. Keith Tribe (Cambridge: MIT Press, 1985), 30.
8 Hayden White, "Historical Discourse and Literary Writing," in *Tropes for the Past: Hayden White and the History/Literature Debate* (Amsterdam: Rodopi, 2006), 25. A resonance between

of possibility. As Campe shows, the emergence of a new scale of probability replaces a model in which possibility and reality are, as depicted by Aristotle, binary alternatives.[9] That is, prior to the seventeenth century possibility and reality were seen as two separate realms or qualities; the emergence of probability theory produced a new understanding of a scaled relationship between them.

In the case of Gottsched the distinction between the possibility that is the realm of the poet and the reality that is the realm of the history-writer is, if not dissolved, at least ambivalent. He represents both genres of writing as trafficking in both possibility and actuality. For instance when Gottsched first compares the poet and the history-writer in the *Dichtkunst*, he writes that the history-writer

> führt seine Helden wohl gar redend ein, und läßt sie oft Dinge sagen, die sie zwar hätten sagen *können*, aber *in der That* niemals gesagt haben: wie wir in griechischen und lateinischen Scribenten häufige Exempel davon vor Augen haben.[10] (emphasis mine)

> depicts his heroes speaking, and often has them say things that they *could* have said, but *in fact* never said; of which we can bring to mind many examples in Greek and Latin writers. (emphasis mine)

Gottsched portrays the history-writer as *inventing* a speech that a hero *might* have delivered. Such a fabrication of a *possible* event would be, in Aristotle's delimitation of possibility and reality, the province of the poet. Gottsched thus contravenes Aristotle's distinction between poet and historiographer as engaged with possibility and reality respectively.[11] To be sure, Gottsched takes Aristotle at his word that representing what a person *might* say is a poetic and not a historiographic operation. He offers a justification for this excessive proximity of the history-writer to the poet: "Let me first say that not everything that

---

Gottsched and White is not in itself surprising, given Gottsched's historical context and White's own indebtedness to principles derived from Vico.

9 Rüdiger Campe, *The Game of Probability: Literature and Calculation from Pascal to Kleist*, trans. Ellwood H. Wiggins Jr. (Stanford: Stanford University Press, 2012).

10 Gottsched, *Dichtkunst*, 148; all translations of Gottsched are my own.

11 In the *Poetics*, Aristotle defines the "universal" element of poetry, as we saw in Chapter One, in the context of comparing it to history: "By a 'universal' I mean the sort of thing that a certain type of man will do or say either probably or necessarily" (1451b8–9). This line is frequently taken as a large-scale claim about the universality of poetry, but in the present context it is noteworthy that Aristotle presents as an example of a poetic fabrication a probable speech, i.e. one which fits what a person *could* have said, whereas Gottsched allows the very same operation of speech-fabrication to the history-writer.

a history-writer does is done *as* a writer of history" (148, emphasis mine). Nonetheless Gottsched's justification acknowledges that the history-writer's inclusion of a merely *possible* speech goes against the canonical connection of poetry to possibility and of history to actuality.

The history-writer may employ poetic operations, in other words, but only *as a borrower*. This inter-occupational practice – if it is acknowledged as such, i.e., if it is recognized that the practice is employed by someone *as a history-writer* – in effect attests to the distinctness of the proper realm of poetry vs. historiography.[12] In the same context, Gottsched authorizes further borrowing, allowing that writers of history may also offer commentary: "A history-writer can of course moralize, and mix political observations into his narratives, as Tacitus and others have done. Is that, however, properly a part of history?" (148). As with the poetic operation of inventing speeches, Gottsched suggests that to the extent that the commentary is political or moral, it is not properly [*eigentlich*] historical. There is, in other words, a proper way to write history that differs from the proper way to compose poetry. Still, the practice of borrowing poetic or moralizing techniques is permitted in writing history. The topic of the proper tools of the history-writer points back to the previous chapter on Aristotle's *Poetics*: precisely within Gottsched's attempt to distinguish the poet and the history-writer, he grants them both the practice that Aristotle – and a whole tradition – embraces for literature, namely the recounting of possible, not actual, events. Gottsched's poet, as we have just seen, fabricates speeches that a *possible* character *really* says in a *possible* world. His history-writer invents a speech that a *real* person *might possibly* have said in the *real* world.

---

[12] We see here Gottsched's own role as a critic, amid what Campe describes as the emergence of the distinct disciplines, when Gottsched reinforces the distinctness of the two kinds of writing while acknowledging that borrowing does happen. See Rüdiger Campe, *Affekt und Ausdruck: Zur Umwandlung der literarischen Rede im 17. und 18. Jahrhundert* (Berlin: De Gruyter, 1990), 56–65). Gottsched acknowledges that his own description of the critic borrows from the Earl of Shaftesbury, among others (Campe, *Affekt und Ausdruck*, 145). P.M. Mitchell, *Johann Christoph Gottsched (1700–1766): Harbinger of German Classicism* (Columbia, S.C.: Camden House, 1995), 29, claims that Gottsched did not necessarily read the German translations of Shaftesbury but instead relied either on French translations or on reviews of Shaftesbury. See also Freier, *Kritische Poetik*. Klaus L. Berghahn has a different view than Campe of how Gottsched is at a crossroads. For Berghahn, Gottsched stands at the point where critique becomes a public matter; Gottsched gives rules for critique to a reading public. There was, in Berghahn's argument, always criticism, but rules produce a reading public that can judge, rather than restricting that judgment to elite critics (Berghahn, "German Literary Theory from Gottsched to Goethe," in *The Eighteenth Century*, vol. 4 of *The Cambridge History of Literary Criticism*, eds. H.B. Nisbet and Claude Rawson [Cambridge: Cambridge University Press, 2005], 522–545).

These forms of invention and borrowing breach, in chiasmic fashion, the difference between possibility and actuality. What is more, the practices allowed to both the poet and the history-writer blur the difference between them.

Gottsched's characterization of the role of possibility in history-writing reveals still other ways in which poetry and history are not easily held apart. Elsewhere in the *Dichtkunst* Gottsched allows that the importation of possibility into actuality is not only a possible but actually a *necessary* practice for the history-writer. He notes that ancient history-writers including Thucydides, Xenophon, and Livius make use of a poetizing imagination [*dichtend{e} Einbildungskraft*], specifically owing to the limitations on available historical material: "How could the aforementioned writers have fulfilled this duty [of producing a history] in attributing speeches to famous people many centuries after their deaths?" (149). In performing their task, history-writers must make use of imagination and possibility, which properly belong to poetry, in order to report actual events when those actual events are long past.[13]

In making a case for a stark difference between the styles of poetic and historical writing, Gottsched also recounts criticisms made of ancient Roman historians Publius Annius Florus and Quintus Curtius Rufus. They were criticized, he notes, for their contrived or "artsy" descriptions [*gekünstelten Beschreibungen*] (149). The condemnation of their writing style indicates that "lively descriptions ultimately belong to the art of poetry" (149). That is, the very fact that these writers of history were condemned for exhibiting a poetic writing style serves as evidence of an essential distinction between poetic and historical writing (359). Gottsched adds that history also utilizes devices that are borrowed from poetry in order to make dry narratives a bit more pleasing (148–149). Such poetic elements are characterized as borrowed and superficial sources of pleasure (221).[14]

---

13 In a related vein, Andres Straßberger, *Johann Christoph Gottsched und die "philosophische" Predigt: Studien zur aufklärerischen Transformation der protestantischen Homiletik im Spannungsfeld von Theologie, Philosophie, Rhetorik und Politik* (Tübingen: Mohr Siebeck, 2010), investigates Gottsched's treatment of the "philosophical" sermon, in which the relationship between philosophical, moral, and theological insights is likewise a question of distinct but connected activity. As Eric Blackall points out, Gottsched's *Erste Gründe der gesammten Weltweisheit* (1734) was a widely consulted philosophy textbook in the eighteenth century (*The Emergence of German as a Literary Language 1700–1775* [Ithaca: Cornell University Press, 1978], 48).

14 The affinity between the poet and the historian appears in a very different context in Gottsched's chapter on periodic style. In the context of asserting that a natural word order is desirable in poetry, Gottsched comes to a moment of infrequent criticism of Aristotle on the grounds of leniency [*Gelindigkeit*]. In *Poetics* 22 (erroneously cited by Gottsched as *Poetics* 23), Aristotle rejects a criticism by a historian against the tragic poets for placing prepositions after

As we saw with the chiasmic relationship between possibility, poetry, reality, and history, in these instances in which Gottsched stresses the difference between poetic and historical style he actually ends up justifying and even recommending their admixture. The artful elements *borrowed* from poetry turn out to be *essential* to history-writing when Gottsched claims that "[the history-writer] *requires* the help of another art, in order to achieve a perfect work" (149, emphasis mine). Historiography is here depicted as dependent upon poetry to render it complete, not accidentally or as a supplement but as a matter of need. Hence insofar as the history-writer *requires* artistic devices to complete his work, the history-poetry distinction seems less definitive than Gottsched otherwise claims. The figure of borrowing upon which Gottsched relies both reinforces the history-poetry genre boundary and hints at its porosity. In the attempt to draw the boundary, and in his blurring of it, we can see Gottsched's own role as that of the critic, who highlights the "as" ("*as* a history-writer") that holds together the practices of poetry and historiography and also holds them apart.[15]

Unruly *Scharfsinnigkeit*

Gottsched portrays reason as definitive in the composition of poetry, even representing reason figurally as the progenitor of the *poetische Schreibart*: "Every line must, so to speak, testify that it had a rational father" (424). This characterization of reason as the father of poetic writing style seems to justify Beiser's description of Gottsched as the "high noon of rationalism."[16] Poetic *thinking*, however, is another matter. Gottsched declares that *Vernunft* cannot univocally be designated the origin of poetic writing: "Reason cannot and should not be it" (426). When Gottsched then defines poetic thinking, he portrays it as a labile connectibility of thoughts:

> Gewisse Geister haben viel Scharfsinnigkeit, wodurch sie gleichsam in einem Augenblicke hundert Eigenschaften von einer Sache, die ihnen vorkömmt, wahrnehmen.

---

their objects; Aristotle defends the poets and praises them for being less prosaic. For Gottsched, Aristotle's praise of the divergence from natural word order offers de facto proof that natural word order exists – in contrast to the claims of Johann Jakob Bodmer, the "Swiss Milton," who suggests that among the Greeks there was no concern for rearranging word order. As is the case with the regimen governing disciplinary differences with respect to historians using poetic elements, the rules of word order are proven in Gottsched's view as long as their breach is acknowledged (Gottsched, *Dichtkunst*, 357).

15 See Campe, *Affekt und Ausdruck*, 3–20.
16 Beiser, *Diotima's Children*, 72.

> Was sie wahrnehmen, das drücket sich, wegen ihrer begierigen Aufmerksamkeit tief in ihr Gedächtniß: und so bald zu anderer Zeit etwas vorfällt, das nur die geringste Ähnlichkeit damit hat; so bringt ihnen die Einbildungskraft dasselbe wiederum hervor. [...] [D]as Gegenwärtige bringt sie aufs Vergangene; das Wirkliche aufs Mögliche, das Empfundene auf alles, was ihm ähnlich ist, oder noch werden kann. (426–427)

> Certain minds are so sharp, that they can perceive as if all at once and in one moment a hundred properties of something that occurs to them. What they perceive with avid attention impresses itself deep into their memory: and so when something happens at another moment that has only the slightest similarity with the first moment, their imagination evokes that first moment again. [...] The present evokes the past, the actual evokes the possible, what has been felt evokes everything that is similar or could become similar.

The *Scharfsinnigkeit* that Gottsched claims is central for poetic thinking produces instantaneous, wildly diverse connections of thoughts to one another. Many different lines of association produce an inexhaustible connectibility: the simultaneous perception of different properties of one thing, person, or event; minimal similarities between things, persons, and events; temporal relations of present, past, and future; and experiences both actual and possible, along with anything that is or could be similar. What connection is *not* thinkable within these parameters? The connections of slight or possible similarity are especially fecund; a dog, for instance, could be seen as slightly or potentially similar to a diamond – for carbon belongs to the composition of both and they both appeared in someone's dreams last week.

Gottsched depicts how such a rampant connectibility of thoughts originates [*entsteht*] in poetic figures themselves, i.e., "out of the frequent and bold metaphors, metonymies and other flowery expressions; out of lively descriptions, apposite short parables, and fiery figures" (429).[17] The faculty of *Scharfsinnigkeit* and the figural associations it evokes yield an unruly possibility of affiliation, association, and relation; it is not a matter of logic, calculation, consequentialism, or abstraction in the mode of reason. This is, one could argue, an embryonic thought of "mediality" in Gottsched, a connectibility without

---

[17] See also Andreas Härter, "Die Rhetorik der 'verblümten Redensarten' in Gottscheds Versuch einer Critischen Dichtkunst," in *Colloquium Helveticum: Cahiers Suisses de Littérature Comparée/Schweizer Hefte für Allgemeine und Vergleichende Literaturwissenschaft/Quaderni Svizzeri di Letteratura Generale e Comparata* 30 (1999): 25–44; and Ulf Gräfe, "Die rationalistische Kontrolle der Metapher in der kritischen Poetik Gottscheds," in *Kommunikative Metaphorik: Die Funktion des literarischen Bildes in der deutschen Literatur von ihren Anfangen bis zur Gegenwart*, ed. Holger A. Pausch (Bonn: Bouvier, 1976), 81–95.

conditions, an "-ability" in the sense that Samuel Weber ascribes to Walter Benjamin's thinking of "-*barkeiten*."[18]

Hence the approach to Gottsched as the "high noon of rationalism" is overly rigid in its claims that rules and rationality constitute Gottsched's only criterion for poetic thinking. Already in Gottsched's own time his reception as an out-and-out rationalist was debatable. Abraham Gotthelf Kästner's treatment of Gottsched in his own century, for instance, shows that it is not anachronistic to consider Gottsched as an aesthetic, poetic thinker rather than as a sheerly rationalist one.[19] More provocatively, we propose that Gottsched represents a proto-post-structuralist concern with possibility, connectibility, and the ungovernability of narrative effect. Gottsched's primary "rule" is that connection rests on a multivalent possibility of affiliation, association, and relation, in which figurality and *Scharfsinnigkeit* are the condition – even the engine – of the production of poetic connections:

> Daher entstehen nun Gleichnisse, verblümte Ausdrücke, Anspielungen, neue Bilder, Beschreibungen, Vergrösserungen, nachdrückliche Redensarten, Folgerungen, Schlüsse, kurz, alles das, was man Einfälle zu nennen pflegt, und die alle insgesammt aus einem solchen lebhaften Kopfe entstehen. Dergleichen Geister nun nennet man poetische Geister. (427)
>
> Thus similes, flowery expressions, allusions, new images, descriptions, amplifications, emphatic ways of speaking, ratiocinations, conclusions, in short, all that one calls "ideas," and overall they originate from such lively noggins. Such minds are called poetic minds.

Gottsched describes here the origins of the associations between thoughts in the poetic mind, but ends up ruling out a thoroughgoing regulation. Rules as such cannot fully characterize the rationality that Gottsched imputes to poetic thinking, except insofar as the "rule" of poetic thinking encompasses *possible* similarity, or *possible* relationship – which makes for a distinctly *unruly* rule.

---

18 See Franz-Josef Deiters, "From Collective Creativity to Authorial Primacy: Gottsched's Reformation of the German Theatre from a Mediological Point of View," in *Collective Creativity: Collaborative Work in the Sciences, Literature and the Arts*, eds. Gerhard Fischer and Florian Vassen (Amsterdam: Rodopi, 2011), 75–86; and Samuel Weber, *Benjamin's –abilities* (Cambridge: Harvard University Press, 2010).
19 See Hermann Stauffer, *Erfindung und Kritik: Rhetorik im Zeichen der Frühaufklärung bei Gottsched und seinen Zeitgenossen* (Frankfurt am Main: Lang, 1997), 232. Stauffer's book reveals an ambivalence that Beiser's "high noon of rationalism" moniker does not acknowledge: Gottsched may be considered an early Enlightenment figure, but Stauffer points out that just six years after Gottsched's death, he is seen in aesthetic, poetic, and anti-Enlightenment terms.

Fabel and possible realities

As opposed to the unruly *Scharfsinnigkeit* that characterizes poetic *thinking*, Gottsched requires poetic *writing* to make its connections comprehensible. Gottsched therefore defines the *poetische Schreibart* as "the presentation of many related thoughts [...] which occurs by means of such sentences and ways of speaking, such that their connection is clearly perceivable" (421). This evocation of the clearly perceptible connection between thoughts marks a constraint on the lability of *Scharfsinnigkeit*. For the high quality of poetry explicitly depends on the way the elements fit together: "Just a chaos of sundry little flourishes patched together does not make for good poetry" (421). Rather, the elements of the poem must go together, must connect in some perceivable way. Hence in the *Dichtkunst* Gottsched examines the ways that words should be connected in poetic writing, that events should be connected, and that thoughts should be connected, precisely because the connections of poetic *thinking* are so indiscriminate.

The problem of proper and adequate connection returns us to Gottsched's gloss on Aristotle and his understanding of *Fabel*. As we have seen in Chapter One, Aristotle defines plot or μῦθος as the "arrangement [σύνθεσιν] of incidents" (1450a5). Gottsched's translation of Aristotle's definition names both a one-to-one connection of events and their overall arrangement, as we have already seen in his definition of *Fabel* as "the composition or connection of elements [*die Zusammensetzung oder Verbindung der Sachen*]" (203).[20] The definition of *Fabel* is in Gottsched's time a bone of contention. He criticizes the moralist Pater Bossu for defining *Fabel* as a combination of a true moral and a "dressing up [*Einkleidung*]" of that moral in a fictional form (203). For Gottsched, in contrast, the issue is primarily connection: "These things [animals, human beings, gods, events, conversations, etc.] must be tied together or connected [*verknüpfet oder verbunden*], so that they compose a coherent whole [*Zusammenhang*], and out of that originates a *Fabel*" (203). *Verknüpfung* and *Zusammenhang* are once again central: the pleasing poem must fit together with the teaching, the matters told

---

[20] In fact the word *Fabel* is used variously by Gottsched to name ancient stories as those of Aesop ("heidnische Fabel," 240), heroic stories ("erdichtet[e] Fabeln von Göttern und Helden," 139), and the pleasant presentation of a moral truth. The range of usage is consonant with the variegated history of the concept. See Wiebke Freytag, "Die Fabel als Allegorie: Zur poetologischen Begriffssprache der Fabeltheorie von der Spätantike bis ins 18. Jahrhundert (II)," *Mittellateinisches Jahrbuch: Internationale Zeitschrift für Mediavistik* 20 (1985): 66–102; Walter Payter Gebhard, "Die Grundlagen der deutschen Fabeldichtung des 16. und 18. Jahrhunderts," in *Fabelforschung*, ed. Peter Hasubek (Darmstadt: Wissenschaftliche Buchgesellschaft, 1983), 298–336; Wolfgang Peter Kayser, "Die Grundlagen der deutschen Fabeldichtung des 16. und 18. Jahrhunderts," in *Fabelforschung*, 79–96.

must be tied, the connections between the poet's thoughts must be clear, and the elements of the poem must cohere. The result of such thoroughgoing connection is that the poem makes up a *Zusammenhang*, a coherent whole.[21]

So far, we have seen Gottsched describe the myriad associations and figural connections between the poet's thoughts and the connections made in the poetic work that render it a coherent whole. Gottsched emphasizes that the coherence of the work and the connections between elements must derive not merely from the words that the poet strings together. Words are to be connected, he declares, only where the matters being written about genuinely *are* connected, where there is a *connexio realis* (160). As Eric Blackall points out, Gottsched objects to the Ciceronian principle of *connexio verbalis* [verbal connection] when it comes to both poetic and oratorical composition.[22] The connections between elements must "really" exist, within the sphere of the poem – which is to say, within the sphere of possibility. Here the difference between Gottsched and Augustin Dornblüth, his contemporary, is instructive. Dornblüth disputed Gottsched's demands for *connexio realis*, claiming that *connexio verbalis* in any case already expresses *connexio realis*, that a *connexio verbalis* serves as a proxy for or an indicator of *connexio realis* and derives its validity from that reality.[23] For Gottsched, on the other hand, *Verbindungswörter*, or

---

[21] The reference to *Zusammenhang* recalls Hans Blumenberg's analysis of how the novel in modernity expresses a new historical concept of reality as a *Kontext*. See Blumenberg, "The Concept of Reality and the Possibility of the Novel," in *New Perspectives in German Literary Criticism: A Collection of Essays*, eds. Richard E. Amacher and Victor Lange, trans. David Henry Wilson (Princeton: Princeton University Press, 1979), 29–48.

[22] Blackall, *Emergence of German*, 164.

[23] Blackall, *Emergence of German*, 184. Dornblüth also emphasized the use of grammatical links to tie periods of a sentence together in oratory, whereas Gottsched believed that the links should be obvious and not require connecting words. Here the question became one of Ciceronian style: Gottsched objected to the periodic sentence, in which the main verb appears at the end, because that sentence structure unifies things that do not properly fit together. In contrast, Dornblüth found the periodic sentence appealing because the closing verb renders the sentence a whole. In each case, Gottsched comes down on the side of *connexio realis*. The periods must *really* belong together, not be forced together by words. This is, in other words, a matter of *Zusammenhang*. As Blackall summarizes Gottsched, "[I]n a well-composed sentence the thought-connections should be self-evident" (185). Hence with respect to oratory Gottsched discourages the use of connecting words in the periods of a sentence, because they make writing forced (*gezwungen*) and even ridiculous. The periods should rather be simple (Gottsched, *Ausführliche Redekunst: Nach Anleitung der alten Griechen und Römer, wie auch der neuern Ausländer, in zweenen Theilen verfasset und itzo mit den Zeugnissen der Alten und Exempeln der grössten deutschen Redner erläutert* [Leipzig: Breitkopf, 1759], 305): "A person speaks and writes much more clearly in this way than when a bunch of thoughts are tied together in a rambling sentence" (*Redekunst*, 305). The point is that there should be less

connecting words, should not be needed to express connections, because the connections should already be present in reality and should only on the basis of that real connection be expressed in the flow of words. Poetic elements must *really* belong together, rather than being artificially joined by words alone in a *connexio verbalis*. Hence Gottsched criticizes authors who are content, as in the case of Lohenstein, when things hang together "more or less" [*einigermaßen*] (449). The connections of the words in the poem instead must follow from "real connections."

Owing to his emphasis on the *connexio realis*, Gottsched also criticizes plays that contain contrived appearances, transformations by gods, and other unrealistic poetic devices (240). A poet must be careful, Gottsched writes, not to put unbelievable things onstage (241). Gottsched's rules require believable plots and reject unbelievable elements. His rules prescribe the real connections of elements, even when the elements are themselves fictitious. For this reason Gottsched disallows the deus ex machina, the magical doings of gods, because these are not real causes. In other words, the principle of *connexio realis* even holds for the fictional work's *possibilities* and its *possible* worlds. Such a *connexio realis* in a poetic *Fabel* in effect constitutes a second-order reality in a possible world, a *Zusammenhang* that is the source of rules for *that* world. The criterion of *connexio realis* for a possible world takes us back to Campe's discussion of the gradation of probability. Gottsched writes, "I therefore believe that I best describe *Fabel* when I say: it is the narration of an event, which is possible under certain circumstances but did not actually happen, in which a useful moral truth lies buried" (204). Borrowing from Christian Wolff's philosophy of possibles he adds, "In philosophical terms one could say that [*Fabel*] is a piece of another world" [*ein Stücke von einer anderen Welt*] (204). Remarkably, Gottsched quotes Wolff, a philosopher, for the purposes of a broader discussion of what we would consider literature:

> Herr Wolf [sic] hat selbst, wo mir recht ist, an einem gewissen Orte seiner philosophischen Schriften gesagt, daß ein wohlgeschriebener Roman [...] für eine Historie aus einer andern Welt anzusehen sey. (204)
>
> Herr Wolf [sic] himself said someplace in his philosophical writings that a well-written novel [...] can be regarded as a history from another world, which seems right to me.

---

thinking required: "The more the listener or reader has to think, the less he can pay attention to each detail, and the worse he is able to understand the speaker or writer" (*Redekunst*, 305).

Gottsched expands this to argue that any *Fabel* can be understood this way – i.e., as a history from another world (204).[24] Just as the connected words in our reality attest to the real connection in the poetic order of possibility, the *Fabel* attests to the possibility of other worlds per se.

Let us note that while the poetic *Fabel* connects possible events to a moral truth, according to Gottsched's reference to Wolff, it also connects another possible world to our real world. In other words, the *Fabel's* mere *existence constitutes* a real connection between reality and possibility. It not merely a possible sequence of events that runs parallel to our reality. Rather, it is a piece of another world, a *possible* world, that nonetheless *really* exists in our world. It is a history, but one that takes place in a realm of the possible. Gottsched's Wolffian reference to the piece of another world is deeply ambiguous: it might be seen to figure a non-integral connection between reality and possibility, or an integral one. The history from another world appears in this world as poetry. Possibility and reality, and thus poetry and history, are themselves connected by *Fabel*.

Clear connections and the constraints of *Zusammenhang*

Let us turn to Gottsched's own comparison between historical and poetic forms of writing in order to examine the connections of possibility and actuality in the explication of a poem.[25] Campe has elucidated this comparison at some length in *Affekt und Ausdruck: Zur Umwandlung der literarischen Rede im 17. und 18. Jahrhundert;* his analysis focuses on the difference it illustrates between Gottsched and Bodmer/Breitinger and on Gottsched's position as a critic.[26] The present analysis, in contrast, concentrates on the principles of connection that Gottsched invents and describes. For having just portrayed the unruliness of *Scharfsinnigkeit* in the connection of thoughts, Gottsched proposes to compare historical and poetic writing. He fabricates a possible historical account that a history-writer might compose to tell of a land that suffers war, hunger, and plague:

> Nachdem der Krieg in dem guten Reiche ein Ende genommen hatte, und die feindlichen Völker abgezogen waren, folgte ein anderes landverderbliches Übel nach. Die verwüsteten Äcker trugen keine Früchte, weil niemand da war, der sie bauen wollte;

---
24 See also Joachim Birke, "Gottscheds Neuorientierung der deutschen Poetik und der Philosophie Wolffs," *Zeitschrift für deutsche Philologie* 85 (1966): 560–575.
25 See also Peter Klaus, "Der gute Herrscher: Literarische Beispiele bei Gottsched und in der Romantik," in *Geschichtlichkeit und Aktualität: Studien zur deutschen Literatur seit der Romantik,* ed. Klaus-Detlef Müller et al. (Tübingen: Niemeyer, 1988), 97–112.
26 Campe, *Affekt und Ausdruck,* 39–44.

und also entstund eine Theurung, die bey dem Armuth nothwendig eine Hungersnoth nach sich ziehen mußte. Auch das war noch nicht alles. Eine pestilenzialische Seuche machte das Elend des geplagten Landes vollkommen, und beraubte es vollends seiner noch übrigen Einwohner. (427)

After the war in the good kingdom had come to an end, and the enemies had withdrawn, another evil followed that ruined the land. The devastated fields bore no fruit, because no one was there who wanted to cultivate them; and thus price inflation occurred, amid such poverty that a famine necessarily followed suit. And that was not all. A pestilential epidemic completed the misery of the plagued land, and deprived it of its last remaining inhabitants.

This is Gottsched's example of the historical way of writing, which meets his own requirement that history be pleasant to read, for it is vivid and suspenseful. Gottsched nonetheless offers the passage as an instance of the *historische Schreibart* because, he writes, it presents its material "clearly and neatly, correctly and delicately, not shamefully but also not showily" (427). "*[D]eutlich und ordentlich, richtig und zierlich, nicht niederträchtig, aber auch nicht prächtig*" – quite a poetic description in its own right, possessing a charming metrical quality and rhyming scheme.

The historical account that Gottsched invents exhibits clear *connexio realis*. The connecting words "because" [*weil*] and "thus" [*also*] do not produce, but instead *reflect*, straightforward and compelling cause-effect relationships between conditions and consequences: no one was there to tend them *and so* the crops did not bear fruit; *because* there was no fruit, there arose a famine. To be sure, the third element, the plague, bears no obvious relationship of consequence to the antecedent war and famine. It seems rather a matter of adding suffering to suffering. It is all the more remarkable that the narrative inserts a connecting phrase, "And that was not all [*Auch das war noch nicht alles*]," which points ahead to the final destruction, in effect instructing us that this is a story of complete devastation. If we review this example carefully, we may note that this fabricated historical account exceeds the requirements that Gottsched establishes for his comparison. His setup reads: "Supposing [*gesetzt*] a history-writer wanted to tell how a land was assailed by three known plagues, war, famine, and epidemic" (427). Gottsched prescribes a simple historical narrative in which war, famine, and plague take place. The fabulated historical account, however, introduces *real* connections between these events: war *brought* poverty, poverty *resulted* in famine, etc. In Gottsched's enactment of the role of a fabulated history-writer, three events are *rendered* a connected narrative, in which plague completes the story of a land's demise. Here the cause/effect connections are obvious and the trajectory of completion is likewise on view.

Now let us turn to Gottsched's example of a poetic rendition of the very same sequence of events, a lengthy passage taken from Amthor:

2 Contingency, Connection, and Possible Worlds

Kaum hatte Mavors Raserey
Den ungeschlachten Durst gekühlet,
Und deine Felder durchgewühlet;
So trat ihm ein Gefährte bei.
Der Mangel ward vom Krieg gebohren;
Weil in der Furchen ödem Grund,
Mehr Blut als warmer Regen stund,
Gieng aller Aecker Zier verlohren.

Dein Elend soll vollkommen seyn!
Zween Feinde hatten dich bestritten:
Noch hast du nicht genug erlitten,
Drum schießt der dritte mit herein.
Morbona bricht durch alle Riegel,
Sie steigt aus einer Todtengruft
Und rühret die vergifte Luft
Durch ihre schwarzgemalte Flügel.

Du wohlgeplagtes Land und Stadt!
Was kann wohl deinen Aengsten gleichen?
Wer zählet die gestreckten Leichen,
Die Mortens Wuth geschlachtet hat?
Du kannst die frechen Seelen lehren,
Was das bedrängte Leben sey:
Und bringst durch tausend Zeugen bey,
Wie sehr die Lust sich kann verkehren. (427–428)

Mars's fury had barely
Cooled its raw thirst
And ransacked your fields
When a companion joined him.
Deprivation was born out of war;
Because in the ground's bleak furrows
Stood more blood than warm rain,
All the splendid fields were lost.

Your misery should be complete!
Two enemies fought against you
But you still have not suffered enough,
So a third joined in as well.
Morbona breaks through all locks,
She rises out of a grave
And stirs the poisoned air
With her blackened wings.

You, afflicted land and city!
What can compare to your worries?

Who counts the rows of corpses
That Death's anger has slaughtered?
You can teach impertinent souls
What a beleaguered life is:
And teach with a thousand witnesses
How desire can go wrong.

Gottsched provides a detailed explanation of the poet's rendition of the events, identifying the similarities and associations that must have produced connections in the poet's mind:

> Dem Poeten sind tausend Dinge eingefallen, daran der Geschichtschreiber nicht gedacht; bey dem Kriege nämlich, der Gott des Krieges, und dessen Blutdurst, imgleichen die Felder, die von einem Heere durchgraben und verderbet worden. Weil die Hungersnoth aus dem Kriege entstanden ist, so fällt es ihm ein, daß die Kinder von ihren Aeltern entstehen: und er braucht also dort das Wort gebohren, welches in ganzes Gleichniß anzeigt. Wenn er die unfruchtbaren Aecker bedenkt: so sieht er, anstatt des Regens, das Blut in den Furchen laufen. Da vorher von Feinden die Rede gewesen, so sieht er, daß auch der Hunger ein Feind des Landes heißen könne; weil er den Kriegsleuten darinn ähnlich ist, daß er Schaden stiftet. Er zählt also schon zween Feinde, und da ihm die Pest noch vor Augen schwebt, davon er reden soll: so macht er sie zum dritten Feinde, weil er eben die Aehnlichkeit daran bemerkt. Die Seuche bringt ihn auf die Morbona: diese läßt er, ihrer Natur gemäß, aus der Gruft steigen, und, weil sie sehr fürchterlich ist, mit schwarzen Flügeln durch die vergiftete Luft fahren. Hierauf sieht er ihre traurige Wirkungen: er entsetzt sich, und bricht in voller Entzückung in eine heftige Anrede und etliche Fragen aus; beschließt aber endlich mit einer Lehre, die aus der Sache fließt, und seine vorige Beschreibung erbaulich macht. Das mag ein Muster einer vollkommen schönen poetischen Schreibart abgeben: Denn 'Omne tulit punctum, qui miscuit utile dulci/Lectorem delectando pariterque monendo'. (428–429)

> A thousand things occur to the poet that would never come to mind for the history-writer; in the case of war, the god of war and his thirst for blood; likewise the fields that are ransacked and laid to waste by an army. Because a famine originates from the war, it occurs to him that children originate from their parents: and thus he uses the word "born," which signifies on the whole a comparison. When he thinks about the barren fields, he sees, instead of rain, blood flowing through the furrows. Since enemies were mentioned earlier, he sees that hunger could also be called an enemy of the land, because it is similar to warriors, in that it inflicts damage. He thus counts two enemies, and since he has an image of a plague in mind, which still must be mentioned, he therefore includes her as a third enemy, because he notices the similarity. The plague brings him to Morbona: he has her, in accord with her nature, rise out of the grave, and because she is very terrifying, he has her fly through the poisoned air with black wings. Afterwards he sees her tragic effects: he becomes horrified and bursts out passionately into a fierce speech and a series of questions; nonetheless he concludes with a lesson that follows from the matter and makes his previous description edifying. This could stand as a model of the perfectly beautiful kind of poetic writing; hence "Whoever mixes the useful with the sweet wins all the points, pleasing and teaching the reader all at once."

We see here Gottsched's narrative reconstruction of the work of *Scharfsinnigkeit*, in which many things occur to the poet that do not occur to the history-writer: the god of war, his bloodthirstiness, the fields in their spoiled state, children born to parents, bloody furrows, etc. The notion of an enemy evokes the relationship of hunger to the land and also evokes the plague as another enemy, because it is similar to these elements. The pollution evokes Morbona and her terrible effects, which enrages the poet and provokes the questions of the last strophe. At the end of his explanation of the connections within the poem, Gottsched also presents a teaching which follows from, and for that reason is directly connected to, what has been said. He articulates the path of associations that renders it a model for a perfect and beautiful way of writing poetry.

As he introduces the Amthor poem, Gottsched declares that the juxtaposition of the two passages illustrates "clear as day [*sonnenklar*]" the difference in thought between the poet and the history-writer: "Now if we compare these to each other, it will be clear as day what the difference is between the types of thought" (428). But what, in fact, does the juxtaposition of passages show us about those differences? Gottsched points to the combination of teaching and pleasantness in the poem, whose significance is emphasized by the closing quotation from Horace. It is not, however, the combination of teaching and pleasantness that makes the poem so different from the historical account. For the fabulated history, which has its own metrical form and descriptive luster, offers a lesson as well, namely that even if your side wins, war may ultimately end in the devastation of your land and the destruction of your people.[27] What is more, the poem's teaching seems incidental to Gottsched's comprehensive elucidation of Amthor's poem. That is, Gottsched seems to linger far longer, and with great vehemence, over the trains of associations than over the teaching, to which he refers only punctually and without passion. Moreover, he fills in the connections of the congeries of elements, connections that are only implicit in the poem – e.g., because he has written how famine originates with war, it occurs to the poet that children originate from their parents; the poet imagines a desolate stretch of land and thinks not of rain but of blood flowing through its furrows. These far-flung connections that Gottsched painstakingly articulates are a more striking point of contrast with the history-writer's account than the moral teaching.

---

27 See also István Gombocz, "'Es ist keine Wissenschaft von seinem Bezirke ganz ausgeschlossen': Johann Christoph Gottsched und das Ideal des aufklärerischen Poeta Doctus," *Daphnis: Zeitschrift für Mittlere Deutsche Literatur* 18, no. 3 (1989): 541–561.

The differences Gottsched's comparison highlights, moreover, do not support what he has stated about the poet vs. the history-writer. Gottsched has ordained that in poetic writing the connection [*Verknüpfung*] between the elements should be entirely clear. But Amthor's poem offers a sequence of elements whose connections are *not* self-evident. The need for Gottsched, as a critic demarcating the differences between history-writing and poetry-writing, to *explain* the connections between the poetic elements stands in sharp contrast to the fabulated history-writer's clear and well-signalled connections. Gottsched's performance as a critical guide to the poetic chain of associations shows that the poetic connections in Amthor – Gottsched's very own example – are precisely *not* obvious. They *require* his critical commentary to elucidate the connections among the poetic elements, although the clarity of real connection is supposed to be the hallmark of poetic writing.

Real connections in fabulated narratives

We have seen in the earlier sections of this chapter that the connection of thoughts within poetry, through *Scharfsinnigkeit* and figurality, is labile. With the consideration of *Fabel* we discover another dimension of the ungovernability of connection that besets Gottsched's rules and distinctions. The connections within a *Fabel* occupy a place in this real world, although they can be understood as coming from another, possible world. They are, to say it another way, possibilities that are represented "into" the reality of this world. The reality of a *connexio realis* proves surprisingly difficult to isolate amid the constellations of possibility, probability, and reality. No wonder Gottsched is worried about believability, defined in terms of connection, in contrast to the deus ex machina. When possibility and reality are no longer separated realms with distinct systems of connection, and instead when a logic of integral probability means that possibility and reality are connected, the unruly possibilities for connection exceed the realm of words and thoughts. The orders of reality and possibility, of *Fabel* and this world, spill over into one another.

Let us return in conclusion to Gottsched's own critical performance between reality and possibility, revisiting how he provides a slow-motion, meticulous explanation of the Amthor poem, explaining what leads the poet's thoughts from one image to another. For Gottsched's fabulated history-writer no such explanation or commentary regarding the connection of thoughts was necessary. But Gottsched, in showing us the difference between poetic and historical writing, himself *tells* us what the poet imagines, what must have occurred to the poet. He narrates what he takes to have been the poet's train of thought, the poet's process of "poetic thinking." In other words, Gottsched provides a *quasi-historical* account, recounting what he himself imagines must have been the poet's sequence of thoughts, such that the connections from thought to thought are rendered clear, as with the historical account the connections from event to event were clear on their own.

Yet of course he *invents* that sequence of connections. Gottsched *fabulates* a *possible* train of Amthor's associations and narrates the sequence *as if* it were real, when it is itself merely a *possible* account of what the poet *may* have thought. Amthor's actual connections of thought are, shall we say, "another world" to Gottsched's connections. Gottsched writes a *possible* historical account of a *possible train* [Zusammenhang] of thought by a *real* poet who constructed a *possible* world. He supplements the wildly associative poetic account with a narration that explains those connections, in order to show that they are really possible, or possibly real, connections. He nonetheless implicitly portrays himself as depicting the *real* poetic Zusammenhang of Amthor's poem.

In other words, only the critical supplementation that Gottsched provides to the poet's passage renders the poetic connections clear. Hence the Aristotelian difference between real and possible events is not an airtight criterion for differentiating history-writing from the art of poetry, not even for Gottsched. He ends up, as we have seen, portraying a far more intimate, and even dependent, relationship of historiography to poetry than he otherwise professes. Historical writing is shown to be shot through with other elements, i.e., liveliness, rich description, poetics, moralizing, fabulation, and so forth. The proper character of historiography does not exclude these other poetic elements, but seems instead to require them. Even more significant are the conundrums of possible and real connections and their intersection in *Fabel* as a piece of another world. Poetry demands the real connection of its possible elements in order to achieve a proper *Fabel* and history demands the borrowing of poetic elements in order to be effective. The real and the possible, the poetic and the historical, are thereby confounded in Gottsched's own prescriptions.

It is perhaps difficult to approach Gottsched outside of a teleological orientation toward Kant, rationalism, and the Enlightenment. As we will see in the next chapter, however, we may also consider the degree to which a concern with connections is central in Kant – e.g. the discontinuity between our cognitive faculties and the noumenal world, the connection of subject and predicate in the form of a judgment, and our connections to other human beings with the shared faculty of rationality. To put it another way, Kant's critical work concerns itself with the ambiguous hinge that holds together various heterogeneous elements of ourselves and the world, hence Hans Vaihinger's 1911 *The*

Philosophy of 'As If'.[28] What is more, for Kant rationality as such admits no direct access but only indirect attestation. In Gottsched, in contrast, the concern for connections in his art of poetry does not require those connections to be rational in a logical sense nor in a sense that Kant would embrace. As the discussion of *connexio realis* has indicated, they are not "as if" connections, but *real* ones, real connections either of the world or of thought. Thus the rational connections made by thought in poetic composition are not logical but are instead a matter of *Scharfsinnigkeit* and the deployment of various devices and principles of connection, including figurality, *connexio realis*, and *Fabel*, by means of which similarities and possibilities verge on and emerge into reality.

---

28 Hans Vaihinger, *Die Philosophie des Als Ob: System der theoretischen, praktischen und religiösen Fiktionen der Menschheit auf Grund eines idealistischen Positivismus* (Leipzig: Meiner, 1922).

Part II: **History: Aesthetic Connection in Historical Knowledge and Historical Composition**

3 *Cognitio historica* between Kant and Meier

It seems obvious that historical knowledge pertains to events of the past, but this was not always the case. In eighteenth-century German epistemology *cognitio historica* [historical knowledge] had nothing directly to do with the past, as Georg F. Meier's *Excerpt from the Doctrine of Reason* [*Auszug aus der Vernunftlehre*, 1752] illuminates. A shortened version of his much longer *Vernunftlehre* [Doctrine of Reason], it defines *cognitio historica*, in German *historische Erkenntnis*, as a matter of empirical knowledge, in opposition to *cognitio rationalis*, or rational knowledge, in German *vernünftige Erkenntnis*, which operates by way of concepts alone. Meier's notion of *historische Erkenntnis* vis-a-vis *vernünftige Erkenntnis* can be understood in the context of Christian Wolff's differentiation between *cognitio philosophica* and *cognitio historica*.[1] According to that schema, *cognitio historica* is in effect a synchronic knowledge of worldly things and facts. Although it is called "historical," it does not refer to the past, nor to memory, temporality, or the connections between events. For the purposes of this book, historical knowledge represents a hinge of sorts between the "piece of another world" described by Gottsched and the development of modern narrative history.

A shift in the philosophical understanding of historical knowledge, from synchronic, empirical knowledge of worldly facts toward the knowledge of the past in narrative form, is subtly visible in Kant's reflections on Meier's *Auszug*. Over four decades (1755–1796) Kant offered over fifty lecture courses on logic that were based on Meier's handbook.[2] In Kant's notes from several years of those lectures, we observe a modification in the notion of *cognitio historica*. While he bases his lectures on Meier's text throughout the "Blomberg logic"

---

[1] See Norbert Hinske, *Zwischen Aufklärung und Vernunftkritik: Studien zum Kantschen Logikcorpus* (Stuttgart: frommann-holzboog, 1998), 29, 112.
[2] Immanuel Kant, *Logik*, vol. 16, *Gesammelte Schriften*, ed. Preußische Akademie der Wissenschaften (Berlin: De Gruyter, 1969). The volume includes the compiled notes that Kant wrote by hand in his copy of Meier's handbook. Meier's own *Auszug aus der Vernunftlehre* is reprinted section by section in footnotes in that volume. Hereafter references to this volume will be abbreviated AA 16. Student notes to Kant's lectures, including what is referred to as the "Blomberg logic," appear in volume 24, no. 1, hereafter abbreviated AA 24.1. Spelling and punctuation conventions differ between the two volumes and within each volume. The Blomberg logic appears in translation in *Lectures on Logic*, ed. and trans. J. Michael Young (Cambridge: Cambridge University Press, 1992). Some of the handwritten notes are translated in Kant, *Notes and Fragments*, ed. Paul Guyer, trans. Curtis Bowman, Paul Guyer, and Frederick Rauscher (Cambridge: Cambridge University Press, 2005). If not otherwise marked, translations of Kant's notes are my own. Translations of Meier are taken from *Excerpt from the Doctrine of Reason*, trans. Aaron Bunch (London: Bloomsbury, 2016).

lectures, Kant's glosses on Meier regarding *cognitio historica*, in contrast to Meier's own explanations, evoke the dimensions of memory, narrative, temporality, and history proper. This is especially the case in Kant's examples, in which he makes reference to narratives of connected events in the past, rather than to isolated empirical facts. This is not to say that Kant thoroughly redefines *cognitio historica* in this context. Indeed, his lectures recapitulate Meier's version of the distinction between historical and rational knowledge. All the same, there is in Kant's digressions – behind Meier's back, we might say – a turn toward connected events, for memory, narrative, and diachrony appear in Kant's explanations of historical knowledge, which in Meier is synchronic, empirical, and not "historical" in a contemporary sense.

The theme of historical narrative is not an obvious focus for an investigation of Kant's uptake of Meier. Kant's lectures on logic, along with his marginal notes to Meier's handbook, are most often studied with an eye to how they lay the ground for the *Critique of the Power of Judgment*.[3] Given the significance of Kant's critical work, it is perhaps difficult *not* to read the lecture transcripts and marginal notes with a view toward the critiques and the preoccupation with subjectivity that arise in their wake. What is more, in considering what "history" means in Kant, the lectures on logic would seem to be one of the last places to look. His "Idea for a Universal History with a Cosmopolitan Purpose" (1784) and his reviews of Herder's *Ideas for a Philosophy of History of Mankind* (1784–1785) are more obvious starting points, as they directly discuss the philosophy of history and the narrative requirements thereof.[4] Similarly, in *Contest of the Faculties* (1798) Kant attempts explicitly to formulate a "predictive history" that might answer the question: "Is the human race constantly improving?"[5] Moreover, Kant's "Conjectures on the Beginnings of Human History" proposes criteria for the distinction between history and fiction.[6] While history

---

3 See, however, Elfriede Conrad, *Kants Logikvorlesungen als neuer Schlüssel zur Architektonik der Kritik der reinen Vernunft: Die Ausarbeitung der Gliederungsentwürfe in den Logikvorlesungen als Auseinandersetzung mit der Tradition* (Stuttgart: frommann-holzboog, 1994), for a detailed argument that the lectures on logic are also essential to understanding the *Critique of Pure Reason*. Riccardo Pozzo, *Kant und das Problem einer Einleitung in die Logik: Ein Beitrag zur Rekonstruktion der historischen Hintergründe von Kants Logik-Kolleg* (Frankfurt am Main: Lang, 1989), offers historical and philosophical background.
4 Immanuel Kant, "Idea for a Universal History with a Cosmopolitan Purpose," in *Political Writings*, ed. Hans Reiss (Cambridge: Cambridge University Press, 1991), 41–53; Kant, "Reviews of Herder's *Ideas on the Philosophy of the History of Mankind*," in *Political Writings*, 201–220.
5 Kant, "Contest of the Faculties," in *Political Writings*, 177–190.
6 Kant, "Conjectures on the Beginnings of Human History," in *Political Writings*, 221. Hence Hayden White famously ties the nineteenth-century turn toward the belief in the objectivity of

per se is directly at stake in these other writings of Kant, as we will explore in the next chapter, nonetheless his notes to Meier's *Auszug aus der Vernunftlehre* demonstrate how *cognitio historica* went from being "an unstructured and in a sense pre-rational 'experience'" to a knowledge of connected past facts and hence a narrative form of historical knowledge.[7]

It is a knotty philological matter to assess Kant's relationship to Meier by way of Kant's lectures on logic, which include direct recapitulation of the Meier handbook and also lengthy expansions, contradictions, and digressions. The provenance of the text is also problematic in numerous respects.[8] The so-called "transcripts" of the lectures are not reliably traceable to any specific auditor who was actually present at the lectures; indeed Hermann Ulrich von Blomberg had already finished his university studies before Kant gave the lectures whose presumed transcripts are called the "Blomberg" transcripts.[9] The presentation of Kant's notes to Meier is also problematic. The volume of the *Akademie* edition devoted to Kant's personal copy of the Meier attempts, in convoluted fashion, to systematize several decades' worth of Kant's accumulated marginal notes. Gerhard Lehmann, who edited the logic lectures for the *Akademie* edition, is said to have done a sloppier job of transcription and editing than his later editorial counterparts Reinhard Brandt and Werner Stark.[10] Hence with the Blomberg logic, and even more so in the case of Kant's notes to Meier, we have "works" that are well past the threshold of disunity and nonidentity: they are philologically problematic, formally disconnected, and intertextually entwined with Meier's *Auszug*. Nonetheless, the notes to Meier and the Blomberg lectures, in particular insofar as Kant criticizes and diverges from Meier, are

---

the historiographical discipline to the maturation of narrativity and specifically to what White calls Kant's "decision to treat historical comprehension as a fiction having distinct moral implications," see White, "The Value of Narrativity in the Representation of Reality," in *The Content of the Form: Narrative Discourse and Historical Representation* (Baltimore: The Johns Hopkins University Press, 1987), 24; White, *Metahistory: The Historical Imagination in Nineteenth-Century Europe* (Baltimore: The Johns Hopkins University Press, 1973), 68.

7 Donald R. Kelley, review of *Cognitio Historica: Die Geschichte als Namengeberin der frühneuzeitlichen Empirie*, by Arno Seifert, and other works. *The Journal of Modern History* 54, no. 2 (1982): 322.

8 For an account of the editorial decisions surrounding the publication of the lecture transcripts see Angelica Nuzzo, *Kant and the Unity of Reason* (Lafayette: Purdue University Press, 2005), 21–22.

9 Hinske, *Zwischen Aufklärung und Vernunftkritik*, xxvii.

10 Steve Naragon, "Kant in the Classroom: Materials to Aid the Study of Kant's Lectures." http://www.manchester.edu/kant/notes/notesIntro.htm.

central for observing the late-eighteenth-century alteration in the meaning of *cognitio historica* from disconnected empirical facts to a unified narrative account of past events.

Temporality and aesthetics in *cognitio historica*

The change in the understanding of *cognitio historica* fits with what Reinhart Koselleck has characterized as a new temporality at the cusp of modernity.[11] Koselleck's *Begriffsgeschichte* of the concept of history itself, as Chapter Eight will discuss in greater detail, famously argues that the notion of a single history emerged around 1800. Prior to that, Koselleck argues, "history" consisted of anecdotes that were recounted in order to offer instruction. The perceived relevance to the present of anecdotes drawn from the past derived from a model of cyclical time in which the past and the present were not essentially distinct from one another. The development of the modern notion of history, in contrast, reflects a new model of linear time.[12] For modern historiography, the past is distinct from the present. It thus demands a narrative form that differs from the recounting of instructive anecdotes. Koselleck formulates the thesis that the emergent concept of history in the eighteenth century owes its existence to an intertwining of poetics and history, insofar as a new, unitary narrative form is central to the development of a modern historiography.[13]

Kant's Blomberg logic lectures exhibit in subtle but definitive form the temporal shift that is Koselleck's hallmark of modernity. For Meier, *cognitio historica* consists of "*nuda facti notitia*," just the bare facts, based on Christian Wolff's formulation.[14] Kant's citations of Meier show that, for Meier, *cognitio historica* is nothing "historical" in our modern sense and does not pertain to

---

[11] Reinhart Koselleck, "Historia Magistra Vitae: The Dissolution of the Topos into the Perspective of a Modernized Historical Process," in *Futures Past: On the Semantics of Historical Time*, trans. Keith Tribe (Cambridge: MIT Press, 1985): 21–38; Koselleck, "'Space of Experience' and 'Horizon of Expectation': Two Historical Categories," in *Futures Past*, 267–288.
[12] Reinhart Koselleck, "Geschichte, Historie," in *Geschichtliche Grundbegriffe: Historisches Lexikon zur politisch-sozialen Sprache in Deutschland*, eds. Otto Brunner, Werner Conze, and Reinhart Koselleck (Stuttgart: Klett, 1975), 647–652.
[13] Koselleck, "Geschichte, Historie," 661.
[14] Arno Seifert, *Cognitio Historica: Die Geschichte als Namengeberin der frühneuzeitlichen Empirie* (Berlin: Duncker & Humblot, 1976), 163–178. Seifert offers a thorough account of *cognitio historica*, but without directly treating its temporalization and narrativization. See also Erich Hassinger, "'Historische' Erkenntnis in der frühen Neuzeit," *Historische Zeitschrift* 226, no. 1 (1978): 89–101.

time, memory, or historical narrative. Where Kant *comments* on Meier, however, he describes *cognitio historica* in ways that evoke history, memory, and narratives of past events.[15] Daniel Fulda's observations concerning the infusion of aesthetic norms into a nascent academic historiography are relevant here. Fulda argues that Enlightenment thought engaged in intense reflection on the problem of the historical representation of diachronically related events, but found itself paralyzed insofar as it understood historical knowledge as systematic and, in effect, synchronic.[16] In order to get past the traditional disdain for history as merely *singularium rerum cognitio* – that is, as a prescientific presentation of individual facts – and in order to attain the level of general scientific knowledge, Enlightenment thought required an organization of knowledge that would allow for a broader integration into a form of narrative.[17] And this is precisely what happens, albeit in miniature, in Kant's reformulation of Meier's *cognitio historica*.

Perhaps most striking is how Kant introduces memory into his expansion of Meier's explanations of *cognitio historica*. In §18 of the *Auszug*, Meier offers a laconic explanation of the difference between *vernünftige Erkenntnis* (rational knowledge) and *historische Erkenntnis (historical knowledge)*. Historical knowledge, also called common knowledge *(cognitio vulgaris)*, pertains to empirical knowledge about specific things:

> Eine jedwede Erkenntniss, in so ferne sie nicht vernünftig ist, wird *eine gemeine* oder *eine historische Erkenntniss* genannt (cognitio vulgaris, historica). Alle Dinge können historisch erkannt werden, und man mag sogar die Gründe derselben erkennen; so lange man

---

15 Relevant here is Albrecht Koschorke's claim, in his "In Praise of the Undefined: Toward a General Theory of Narrativity," lecture at the Wissenschaftskolleg zu Berlin, 2011, that "stories fill the breach when typification fails," and thus narrative shares certain features with concepts and even serves as an epistemic category.

16 Daniel Fulda, *Wissenschaft aus Kunst: Die Entstehung der modernen deutschen Geschichtsschreibung 1760–1860* (Berlin: De Gruyter, 1996), 22–28. Fulda portrays aestheticization as the *sine qua non* of the advent of modern historiography – i.e. arguing that aesthetics, even more than science, gives modern historiography its method. This is not quite the same as Hayden White's account of the maturation of historical narrativity; Fulda accepts White's claims for a metahistorical framework governing individual histories, but denies the tetradic delimitation provided in White's early work of those metahistorical narratives. Fulda's description of the significance of aestheticization to the philosophy of history forms a stark contrast to Karl Ameriks's argument that in Kant's time, aesthetics served a superficial rhetorical function rather than a more deeply structuring one; see Karl Ameriks, *Kant and the Historical Turn: Philosophy as Critical Interpretation* (Oxford: Oxford University Press, 2006), 283.

17 Fulda, *Wissenschaft aus Kunst*, 156–157; see also Fulda, "Literary Criticism and Historical Science: The Textuality of History in the Age of Goethe – and Beyond," in *The Discovery of Historicity in German Idealism and Historism*, ed. Peter Koslowski (Berlin: De Gruyter, 2005), 112–133.

> den Zusammenhang der Folgen mit ihren Gründen nicht deutlich einsieht, so lange hat man nur eine bloss historische Erkenntniss. (AA 16:94, *sic* emphasis)

> Every knowledge, insofar as it is not rational, is called *common* or *historical knowledge* (cognitio vulgaris, historica). All things can be known historically, and one may even know their grounds; so long as one does not distinctly understand the connection of the consequences with their grounds, one has only a merely historical knowledge. (*Excerpt*, 6, translation modified)

For Meier, following Wolff, historical knowledge is distinct from rational knowledge – but may constitute a precursor to rational knowledge, as we will see below. Historical knowledge knows individual empirical things, but not in terms of their *connection*, which Meier defines as the province of rational knowledge. This definition of rational knowledge conforms with Wolff's understanding of philosophical knowledge as the form of knowledge that connects events to their causes. It hence indirectly involves temporality, i.e., insofar as the relationship of cause and effect is ascribed to events that take place in a sequence.

Meier's description of historical knowledge indicates that it is not temporal; not a matter of a span of time, events, or memory; and does not deal with causal connection. Historical knowledge instead constitutes a cumulative knowledge of individual facts. In his lectures, however, Kant "summarizes" Meier to quite different effect. Where Kant's lectures define *vernünftige Erkenntnis* and *historische Erkenntnis,* he supplements Meier's understanding of that difference with reference to memory and the task of memorizing. Kant's gloss in the Blomberg transcript of the Meier passage just quoted runs as follows:

> Eine Erkenntniß, die nicht vernünftig ist, wo ich nicht Vernunft, *sonderen nur Gedächtniß* brauche, wobey ich nicht sehe, wie etwas aus seinem Grunde herkommet, ist eine Historische Erkenntniß. (AA 24.1:47, emphasis mine)

> A knowledge that is not rational, where I do not need reason but *only memory*, in which case I do not see how something derives from its ground, is a historical knowledge. (*Lectures on Logic*, 33–34, emphasis mine, translation modified)

We observe here that Kant *adds* to Meier's explanation of historical knowledge a reference to a faculty of memory. For Meier, historical knowledge is a knowledge of individual things and, as we have seen, it admits of at most unclear ("*nicht deutlich*") connections with their causes or grounds. A clear connection between things and their grounds would belong instead to the sphere of rational knowledge. For Kant, on the other hand, historical knowledge is a matter specifically of *memory* – although memory makes no appearance in Meier's account. To summarize: for both Kant and Meier historical knowledge does not connect events or things to their reasons, which would be the task of rational knowledge, but as opposed to Meier, Kant's historical knowledge is defined as a matter of memory, of connection to the past.

A closer look at the differences between Meier's and Kant's formulations in these passages of the relationship to grounds reveals in another, quite subtle, way how temporality emerges and events are connected in Kant's gloss on Meier. That is, Meier refers to the relationship of things to their grounds as a *Zusammenhang* ("so long as one does not distinctly understand the connection [*Zusammenhang*] of the consequences with their grounds, one has only a merely historical knowledge"). Kant, however, defines that relationship as "how something derives from its ground [*wie etwas aus seinem Grunde herkommet*]," and thereby suggests, with the use of the verb *herkommet* and even the adverbial *wie*, the phenomenon of temporal flow as determining that relationship. The relationship of phenomenon to ground is for Kant one of *derivation*, which involves a temporal process, whereas for Meier the relationship is one of *Zusammenhang*, which implies a structural, synchronic connection.

Kant likewise inserts memory and past history into his reflections on Meier's *historische Erkenntnis* in a marginal note to Meier's §18 – the same section in which, as we have seen, Meier defines historical knowledge in contrast to rational knowledge. Kant expands on Meier as follows:

> Dieses Erkentniß, wenn es gleich bloß historisch ist, ist darum nicht unvernünftig. Denn Unvernunft ist nicht allein ein Mangel der Anwendung der vernunft, sondern eine [Mangel] anwendung der Krafte gegen die Vernunft. E.g., Wenn ich die *Historie* bloß nach Bildern *lerne* [Ein]. (AA 16:95; editorial brackets reproduce the text of the edition, emphasis mine)

> This knowledge, when it is merely historical, is not for that reason irrational. For irrationality is not just a failure to apply reason, but an [lacuna] application of powers contrary to reason. E.g., If *I learn history* merely from pictures. (translation and emphasis mine)

The heart of this partially corrupt note, dated by the editors to 1752, appears to refer to the distinction between *lacking* reason and *being contrary to* reason. In this context, Kant mulls over the relationship of historical knowledge to rationality and considers whether they are mutually exclusive. What makes this note significant for the present argument, however, is that Kant invokes the example of *learning history*. Learning history per se has nothing to do with the historical knowledge that is being explicated in §18 of Meier's *Auszug*, as we have seen in the Meier passage already quoted. And precisely for that reason

Kant's illustration is remarkable: he takes the scenario of *learning actual history* as his example of the historical knowledge that for Meier pertains to empirical, one-by-one phenomena. What is more, the act of learning history is itself a temporal process, a matter of memory, and it results in connections. Hence Kant's example involves the learning process, past history, and memory – three aspects of temporalization. Kant's gloss on historical knowledge in Meier illustrates it in terms of history, memory, and narrative rather than in terms of a synchronous knowledge of empirical facts. In the Blomberg logic transcripts, Kant refers again to memory and the connection to the past while offering another example of how historical knowledge is not necessarily or intrinsically opposed to rational knowledge:

> Wenn die Vernunft aber bey einer Vorstellung nicht besonders adhibiret wird, so ist dieselbige keine vernünftige Erkenntniß ist deshalb auch keine unvernünftige. Z.E. wenn ich weiß in was für Landen Europa eingetheilt ist, so gebrauche ich dazu nicht Vernunft, sondern nur *Gedächtniß*. was aber keine vernünftige ist, ist noch nicht unvernünftig. (AA 24.1:47, emphasis mine)

> When, in a representation, reason is not applied particularly, then this is not a rational knowledge[, but] it is not on that account irrational either. E.g., when I know into what countries Europe is divided, I do not use reason for this but only *memory*, and on this account it is not a rational knowledge. But what is not a rational knowledge is not then irrational, but rather this holds only for that which is against reason. (*Lectures on Logic*, 33; *sic* editorial brackets, translation modified, emphasis mine)

As with the previous example taken from Kant's marginal notes to Meier, this explanation of historical knowledge explicitly defines it in terms of memory and also alludes to the temporal process of learning, suggesting the scenario of a geography lesson.[18] The knowledge of the different European lands typifies a synchronic matter of spatial-geographical facts, but Kant directly ties that knowledge to the temporal process of memorization.

In yet another example of how Kant inserts memory and hence connection to past events into his account of Meier's synchronic historical knowledge, we see that in §55–64 of the *Auszug* Meier considers how learned knowledge increases and can become extensive and complete [*ausführlich und vollständig*].[19] Meier, however, represents such an expansion of learned knowledge as a quantitative increase in §58–59:

---

18 Seifert, *Cognitio Historica*, 192–195, emphasizes the role of geography in historical knowledge.
19 AA 16:202; *Excerpt*, 13.

> Weil, durch die Erweiterung der gelehrten Erkenntniss, die Erkenntnisskräfte zu gleicher Zeit fähiger gemacht werden; so hat man nicht zu besorgen, dass man durch die beständige Erweiterung der gelehrten Erkenntniss seinen Kopf überladen werde, wenn man nur bei dieser Beschäftigung die Grenzen des gelehrten Horizonts nicht überschreitet. [§59] Ob gleich die Kunst lang und das menschliche Leben kurz ist, so muss uns diese Betrachtung vielmehr anreizen, mit der gehörigen Eilfertigkeit so viel zu lernen als möglich ist. (AA 16:203)
>
> Because, through the broadening of learned knowledge, the powers of knowledge are at the same time made more capable, one need not worry that through the constant broadening of learned knowledge one's head will be overburdened, if only in this pursuit one does not overstep the bounds of the learned horizon. [§59] Although art is long and human life is short, this observation must spur us to learn as much as possible with proper haste. (*Excerpt*, 14, translation modified)

For Meier, what is at stake here is a quantitative increase in learned knowledge, represented as a loading or cluttering one's head with knowledge; the metaphor is primarily spatial. In his lectures, Kant offers a comment that appears to be a gloss on Meier, reiterating the Hippocratean contrast between art and life. In Kant's marginal comment, however, gains in knowledge are represented in terms of an expansion of *memory* of past events, rather than as a quantitative accumulation of sheer facts. What is more, Kant takes *history* as his example of such a gain in knowledge:

> Hippokrates sagt weise: Ars longa, vita brevis. [...] Viele Wißenschaften sind von der Art, daß mit der Dauer der Zeit durch ihren Umfang die Menschlichen Fähigkeiten noch überschritten werden. So ist z.E. die *Historie* schon jezt gar sehr weitläuftig *mit der Zeit* und deren Länge nun geschieht noch immer mehr. es tragen sich noch immer mehr und mehr Begebenheiten in der Welt zu. Diese kommen alle zur *Geschichte* zu, und diese Wißenschaft wird endlich also weitläuftig werden, und zunehmen, daß endlich unser *Gedächtniß* lange nicht groß genung seyn wird. da es jezt schon so schwer ist. (AA 24.1:74, emphasis mine)
>
> Hippocrates wisely says: *ars longa, vita brevis.* [...] Many sciences are of such a kind that, with the *passage of time*, human capacities will be overstepped by their extent. Thus *history*, e.g., is already very extensive[;] with *time* and its duration more and more is always happening. More and more events are always occurring in the world. These all add to *history*, and this science finally will thus become extensive and grow, so that our memory will finally be far too small. For it is already quite hard now. (*Lectures on Logic,* 56, emphasis mine)

Here Kant explicitly attributes *to time* the reason that human faculties can be overwhelmed by the growth of knowledge, i.e., insofar as the various disciplines expand their knowledge over time. This is most obviously the case with regard to history because more and more events are always happening, and for that reason our memory is inadequate to maintain a knowledge of ever-increasing history.

Kant defines memory as "too small" to accommodate the perpetual growth of history, a spatial figure in the context of temporal accumulation.

Given the difference between Meier's representation of memory as a quantitative store and Kant's representation of memory largely as a temporalized accumulation, it is not surprising that forgetting is, in similar fashion, represented quantitatively by Meier and temporally by Kant. For Meier, just as learned knowledge is a matter of increasing the quantity of our knowledge, forgetting consists of the corresponding decrease of knowledge to make room for new learning, as he writes in §61:

> Da es natürlich nothwendig ist, dass wir Menschen vieles vergessen; so muss man eben deswegen die gelehrte Erkenntniss sehr erweitern, damit man viel vergessen und dem ohnerachtet noch viel behalten könne. (AA 16.1:204)

> Since naturally it is necessary that we human beings forget much, just for that reason one must broaden learned knowledge very much, so that one can forget much and nevertheless still retain much. (*Excerpt*, 14, translation modified)

For Meier, as this quotation shows, forgetting is about a reduction in a stored quantity of knowledge in order to maintain a quantitative equilibrium with what is learned and retained. For Kant, in contrast, as we see in the previous quotation responding to the Hippocratic reference, the origin of forgetting is instead the sheer, ineluctable occurrence of more and more events owing to the passage of time. The happening of more and more events certainly complicates empirical knowledge because it means, as Meier acknowledges, that there is an ever-increasing quantity of facts. But as we saw in the Hippocrates comment by Kant on Meier, what is significant is that this complicates our knowledge of *history* because there are more and more events to know about if we are to know history – "*nun geschieht noch immer mehr* [more is always happening]." Hence for Kant the historical knowledge of events connected over time occupies the role of what for Meier is a quantitative epistemological accumulation.[20] The continuity of time, rather than a surplus inventory of memories, results in forgetting.

Let us examine one other attempt by Kant to differentiate *historische* and *vernünftige Erkenntnis*. Here is another instance in which Kant implicitly renders *historische Erkenntnis* more "historical" in the modern sense:

---

[20] Meier recognizes this accumulation of history much later in his handbook when he writes, "With time every history becomes too brief" (*Excerpt*, 122). This remark, however, appears in the context of a discussion of what belongs in a book of history, rather than in the context of the problem that temporalization poses to our knowledge per se.

> Eine Erkenntniß, die nicht vernünftig ist, ist die Historische. Z.E. wenn jemand große Kenntniß der *Natur-Geschichte* hat, und er sie nicht im Zusammenhange mit ihren Gründen erkennet, so hat er keine vernünftige, sondern nur eine Historische Erkenntniß." (AA 24.1:48, emphasis mine)
>
> Historical knowledge is a knowledge that is not rational. E.g., when someone has much acquaintance with *natural history*, and he does not know this in connection with its grounds, then he does not have rational but only historical knowledge. (*Lectures on Logic*, 34, emphasis mine, translation modified)

Once again, Kant tackles the relationship between historical and rational knowledge. In some respects Kant reverts to a more synchronically-oriented understanding of the difference between historical and rational knowledge, namely in using the vocabulary of *Zusammenhang* and *Gründe*. On the other hand, Kant's example evokes precisely natural *history* as a model of *historische Erkenntnis*. The designation *Naturgeschichte* [natural history] of course has a semiotic history of its own.[21] As referenced by Kant in response to Meier, however, *Naturgeschichte* can be seen as a nexus between synchronic and diachronic – or between epistemological and historical – understandings of history. That is, on the one hand *Naturgeschichte* refers to an empirical acquaintance with facts known about nature. On the other hand, *Naturgeschichte* is also a *historical* phenomenon, in two separate ways: it is an emerging scholarly discipline in the eighteenth century, but also *Naturgeschichte* is a type of *history*, i.e., as a discipline it relates a diachronic narrative of the development of nature to its present state. For this reason, Kant's example of *Naturgeschichte* evokes both Meier's synchronic notion of historical knowledge as an accretion of facts and Kant's temporalized notion of historical knowledge as pertaining to connected events.

Perfect and beautiful knowledge

Kant's glosses on Meier with respect to the impossibly large range of historical knowledge occur within the context of a question concerning whether historical knowledge can be complete. This in turn relates to the question, in Kant's reading of Meier, of whether historical knowledge can be beautiful. This seemingly

---

[21] See John G.T. Anderson, *Deep Things Out of Darkness: A History of Natural History* (Berkeley: University of California Press, 2013), for a broad treatment of the history of natural history. See Wolf Lepenies, *Das Ende der Naturgeschichte: Wandel kultureller Selbstverständlichkeiten in der Wissenschaft des 18. und 19. Jahrhunderts* (Munich: Hanser, 1976), for a discussion of temporalization and natural history, and especially 37–38 for a study of how this operates in Kant.

odd formulation actually belongs to the nascent aesthetic philosophy of the eighteenth century. For Wolff and Baumgarten after him, in a way of thinking about perfection or completeness that goes back to antiquity, beauty is tied to *Vollkommenheit*, or completeness. Meier's *Auszug* §19 states that "A more perfect historical knowledge is *a beautiful knowledge* (cognitio pulcra, aesthetica)" (*Excerpt*, 6, translation modified). For Meier, what is at stake here is a relationship between beauty, perfection, and historical knowledge. Kant will reject Meier's portrayal of that relationship in his comments, but before turning to that rejection, let us investigate Meier's position more closely.

First let us note that Meier defines historical knowledge as a *precondition* for rational knowledge, but not as its *precursor*. That is, historical knowledge can never develop into rational knowledge, even if the historical knowledge is perfectly complete [*vollkommen*], which for Meier would mean that it is beautiful. Hence Meier claims in §20,

> Obgleich die historische Erkenntniss von der vernünftigen sehr unterschieden ist, dergestalt, dass die allerschönste historische Erkenntniss nicht einmal eine vernünftige Erkenntniss genennet zu werden verdient, so ist doch jene zu dieser unentbehrlich, indem ein Mensch keine vernünftige Erkenntniss von einer Sache erlangen kann, wenn er nicht vorher eine historische Erkenntniss von derselben besitzt. (AA 16:100–101)

> Although historical knowledge is very distinct from rational knowledge, to such an extent that the most beautiful historical knowledge does not even deserve to be called a rational knowledge; nonetheless the former is indispensable to the latter, since a human being cannot attain rational knowledge of a matter if he does not first possess a historical knowledge of it. (*Excerpt*, 7, translation modified)

For Meier, beautiful historical knowledge, as we have seen in the previous section, does not pertain to the past. It is a matter of a complete, but sheerly empirical, knowledge of something; here Meier adds that historical knowledge is an indispensable *prerequisite* for rational knowledge. His reference to beauty in this context indicates that Meier holds to the aesthetic rationalist position that knowledge and pleasure are connected by way of the intuition of perfection, or *intuitio perfectionis*.[22] This understanding of beauty as the contemplation of perfection derives from Leibniz and Wolff, but Meier conceives of it as a cognitive mode, based on Baumgarten's description of such contemplation in which one observes the harmony of many combined into one.

---

[22] See Frederick C. Beiser, *Diotima's Children: German Aesthetic Rationalism from Leibniz to Lessing* (Oxford: Oxford University Press, 2009), 9, 135, 137.

Kant's comments on Meier's depiction of the relationship between beauty and perfection in knowledge further illuminate his divergences from Meier. In Kant's discussion of Meier §19–20, he follows Meier in declaring that historical knowledge is a matter of acquaintance with something in the empirical sense. Nonetheless, after he quotes, with slight modifications, Meier's statement that a perfect or complete historical knowledge is also beautiful, Kant then rejects Meier on this very point:

> Eine Vollkommene Historische Erkenntniß sagt der Autor [Meier–KF] [...] ist eine Cognitio pulchra sive aesthetica. es ist aber falsch, daß er schön, und Aesthetisch vor einerley hält, denn zur Aesthetic gehöret nicht nur das Schöne, sonderen auch das Erhabene. Eine Historische Erkenntniß ist eine solche die nicht Vernünftig ist. wenn dahero der Autor von einer vollkommenen Historischen Erkenntniß redet, und sie vor eine aesthetische hält, so kann er nichts anderes darunter verstehen, als eine Erkenntniß, die ob sie gleich nicht vernünftig ist, doch noch eine Vollkommenheit haben kann. Dadurch aber hat er sich sehr schlecht erkläret. (AA 24.1:47–48)

> A perfect historical knowledge, says the author [Meier] [...] is a *cognitio pulchra sive aesthetica*. It is, wrong, however, for [Meier] to take beautiful and aesthetic to be the same, for aesthetics includes not only the beautiful but also the sublime. A historical knowledge is one that is not rational. When the author speaks of a perfect historical knowledge, therefore, and takes it to be an aesthetic knowledge, he can understand by this nothing but a knowledge which, though it is not rational, can still have a perfection. But he thereby explained himself very badly. (*Lectures on Logic*, 34, translation modified)

As we see in these lines, Kant objects to Meier's apposition of *pulcra* and *aesthetica*. For Kant, these are not synonyms because the sublime belongs to the realm of aesthetics along with the beautiful. This is of course significant for Kant's later considerations of aesthetics. More important for the present argument is that Kant goes on to criticize Meier's association of beauty and perfection with historical knowledge. Kant's depiction of this constellation is more complex than Meier allows. Moreover, in Kant's eyes, Meier does not explain clearly the nature of a *cognitio aesthetica*. Meier's explanation is therefore an empty one, for which reason Kant adds, "This amounts to my wanting to say: A periwig is that which, even if it is not exactly a hat, nor a cap, can nevertheless be placed on the head just as well."[23]

---

[23] "Dieses ist eben so viel, als wenn ich sagen wolte: die Perruque ist das, was, wenn es gleich nicht ein Hut, noch eine Mütze ist, dennoch eben so wohl auf dem Kopf gesetzet werden kann" (Kant, *Lectures on Logic*, 34; AA 24.1:48).

Beiser, lamenting the waning of rationalist aesthetics, portrays Kant as conducting a "polemic against perfection," but we see here that Kant does not so much reject perfection as dissociate it from beauty.[24] Kant's point is to contrast Meier's claim for a perfect and beautiful historical knowledge, which Kant views as "empty," to Kant's own explanation of the aesthetic perfection of a knowledge based on its effect on *feeling*. The topic of feeling versus perfection is an enormously important one for the genealogy of aesthetics; for our purpose it is most important to note that perfection, for Kant, is simply the wrong criterion for beautiful historical knowledge. Kant's counterexamples speak against Meier's suggestion that historical knowledge is more beautiful the more perfect or complete it is. They imply instead that feeling, rather than perfection, is the decisive criterion for aesthetic knowledge.

With respect to this chapter's emphasis on narrative in historical knowledge, what is most significant is that Kant's examples of beautiful historical knowledge reflect a thoroughly *temporalized* understanding of historical knowledge, in which the sequence of connected events is primary. Kant's example of Homer's *Odyssey* is particularly noteworthy in this context, namely in response to §18–19 of Meier's *Auszug*. Kant recapitulates and then rejects Meier's claim that a perfect historical knowledge will necessarily be beautiful. Kant takes as his example what we know from Homer's narrative about Odysseus's journeys:

> Der Autor [Meier] halt eine jede vollkommene Historische Erkenntniß vor eine schöne. es kann aber eine [vollkommene – KF] Erkenntniß Historisch seyn, ohne daß sie schön ist. wir können aus dem Homero alles Mythologische auch alle Reisen des Ulysses, etc. und mehr wißen, so daß wir eine recht vollkommene Historische Erkenntnis haben, und dennoch wird sie nicht schön seyn, wenn sie keine wirckung auf unser Gefühl hat, wenn wir alles kalt ansehen, und nichts dabey empfinden. (AA 24.1:48)

> The author [Meier] holds every perfect historical knowledge to be beautiful. But a knowledge can be historical without being beautiful. We can know from Homer everything mythological, even all the travels of Ulysses, etc., and more, so that we have a quite perfect historical knowledge[;] and nevertheless it will not be beautiful if it has no effect on our feeling, if we regard everything coldly and we sense nothing along with it. (*Lectures on Logic*, 34)

Meier has stated that a perfect historical knowledge, i.e., a full account of some worldly phenomenon, is beautiful. As a counterexample Kant takes Homer's *Odyssey*, which, Kant grants, offers a quite complete historical knowledge ("*eine recht vollkommene historische Erkenntnis*") of the mythology of Odysseus

---

24 Beiser, *Diotima's Children*, 18.

and his travels. Kant argues, however, that it is not because it gives a *complete* account of Odysseus' travels that Homer's *Odyssey* is so beautiful. Kant introduces instead the notion that the effect something has on our *feeling* is what makes it beautiful. Here the connection of the logic lectures to the *Critique of the Power of Judgment* is obvious, for Kant's point is that not completeness but instead feeling is the proper criterion for beauty.[25]

With respect to the change in *cognitio historica* between Kant and Meier, what is also relevant is that Homer's *Odyssey* exemplifies a historical knowledge that is itself a *narrative of events*, and specifically a narrative of a "historical" – albeit mythological – past. The "quite perfect historical knowledge" that Kant says we take from the *Odyssey* is, moreover, doubly diachronic. First, it involves the reader's knowledge of the text of the *Odyssey*, which entails the temporal process of reading, and second, the *Odyssey* is itself a diachronic narrative of Odysseus's travels. For Kant, therefore, the temporal process of reading produces the feeling that evokes a judgment of beauty. Such a judgment is tied to a narrative of a mythologized past, a historical knowledge in Kant's sense, in the case of the *Odyssey*. Meier's perfect knowledge is synchronic; Kant's knowledge of the beautiful, in the case of the *Odyssey* at least, involves several levels of diachrony.

The divergence in the concept of *cognitio historica* between beauty and perfection, and between synchronic vs. diachronic experience, becomes even clearer when we take into account the background to Meier's notion of beauty. Baumgarten's theory of pleasure and perfection, from which Meier derives his statements, is based on Wolff's brief account of *pulchritudo* in the *Psychologica Empirica*, where it refers to the intuition of perfection.[26] Baumgarten's *pulchritudo* is a *synchronic* criterion for an objective completeness rather than a diachronic one and Meier's *pulcra sive aesthetica* is likewise synchronic. For Kant, in contrast, the temporality of cause and effect is in play, which entails the connection between events. What makes Homer's historical knowledge beautiful for Kant is not its perfection, but instead its effect on our feelings. That feeling depends upon a temporal unfolding – the unfolding of the reading process and the narrative unfolding of Odysseus's travels.

---

[25] This corresponds to Beiser's claim that with Kant, the pinnacle of aesthetic rationalism is lost because the criterion of perfection of the object is abandoned, see Beiser, *Diotima's Children*, 17.
[26] Quoted in Beiser, *Diotima's Children*, 145.

Narrative knowledge and the temporality of history

These passages in the Blomberg logic and in Kant's notes to Meier reflect the emergence of a narrativized, memory-based historical knowledge in which events are connected, as opposed to Meier's earlier understanding of *cognitio historica* as a synchronic accretion of empirical facts. Kant's reformulation of historical knowledge demonstrates a substantial widening of its scope. Historical knowledge in Kant's sense temporalizes Meier's synchronic *cognitio historica*, but further extends to the very emergence of history as an object of knowledge. Hence this analysis pertains to wider discussions of philosophical and literary scholarship on Kant, the philosophy of history, and the development of historiography.

Kant's modification of the use of the term *historische Erkenntnis* can be illuminated by Koselleck's conceptual history, or *Begriffsgeschichte*, of the concept of history. As Chapter Eight will further explore, Koselleck famously contends that in the eighteenth century *Historie* was conceived as a set of exempla that highlight a constancy of human nature. At the end of the eighteenth century, however, the concept of history was on the verge of a shift toward a unitary *Geschichte* of events recounted in a linear narrative.[27] As Koselleck shows, the term *historiae* evolves from a plural to the singular noun *die Geschichte*. In this turn, the boundary with poetics becomes porous, such that history comes to be bound by rules of unity borrowed from literary spheres. It becomes a "single history," for instance with Wilhelm von Humboldt's development of the idea of "history in general."[28] Koselleck's account of the evolution of the concept of history suggests that modern *Geschichte* embodies a convergence of event and representation – which in its completion heralds the arrival of idealist philosophy.[29]

Kant's early notes and lectures on logic combine rationalist, aesthetic, and historical concerns. They bespeak a strand of thought that remains closer to what David Wellbery has described as a "fifty-year reign of Enlightenment aesthetics [...] [w]edged [...] between the vast cultural formations of rhetoric on one hand and idealist hermeneutics on the other."[30] In inquiring into the

---

[27] Koselleck, "Historia Magistra Vitae," 23, 27.
[28] Koselleck, "Historia Magistra Vitae," 30.
[29] Koselleck, "Historia Magistra Vitae," 27–28.
[30] David Wellbery, "Aesthetic Media: The Structure of Aesthetic Theory before Kant," in *Regimes of Description: In the Archive of the Eighteenth Century*, eds. John Bender and Michael Marrinan (Stanford: Stanford University Press, 2005), 211. See Rüdiger Campe, *Affekt und Ausdruck: Zur Umwandlung der literarischen Rede im 17. und 18. Jahrhundert* (Tübingen: Niemeyer, 1990), for an exhaustive account of the transition in the eighteenth century from rhetoric to hermeneutics.

temporalization and narrativization of *cognitio historica*, we have observed an admittedly obscure germ of the modern philosophy of history. Within Kant's lectures and in this moment that sets up Kant's uptake of Baumgarten's aesthetics, temporality is introduced into knowledge in general when it pertains to worldly, "historical" matters. Kant's introduction of temporality into knowledge in effect recasts narrative as a temporalized epistemic category. The question of how events are connected to one another displaces an emphasis on empirical knowledge of individual facts. Prior to Kant's critical formulations regarding the play of imagination and understanding, we glimpse here how narrative inhabits that act of knowledge and history is a privileged example of this. Kant's glosses on and marginal notes to Meier demonstrate a turning point in the uptake of narrative and history into knowledge itself.

4 "On the Wings of Imagination": Wholeness and Spontaneity in Kant's Philosophy of Universal History

Approximately ten years after Kant gave the lectures that make up the Blomberg logic, he undertook a more focused examination of the criteria for a properly narrative history. His essays from the 1780s are far more famous than the logic lectures and yet they can also be seen as the culmination of the question of historical knowledge, understood as "knowledge of the past." In "On the Idea of Universal History with a Cosmopolitan Purpose" (1784), Kant asks how the history of humankind, "this senseless course of human events," can be represented as forming a unitary whole under the governance of a *Leitfaden*, or guiding thread.[1] One obstacle to this narrative project, as he notes in "Conjectures on the Beginning of Human History" (1786), is the absence of records of our earliest times. Kant proposes to fill this lacuna in the account of human history by means of a conjectural "journey," to be made "on the wings of imagination."[2] In his prescriptions for portraying less distant events, in contrast, Kant is more circumspect: it is permissible, he writes, to intersperse [*einstreuen*] conjectures in a historical account in order to fill gaps. The account, however, must not be based solely on conjectures, lest it amount to "a work of fiction [*eine bloße Erdichtung*]," rather than of history. Indeed, excessive use of conjecture "would seem little better than drawing up a plan for a novel."[3]

---

[1] Immanuel Kant, "Idea for a Universal History with a Cosmopolitan Purpose," in *Kant: Political Writings*, ed. Hans Reiss, trans. H.B. Nisbet (Cambridge: Cambridge University Press, 1992), 42. The theme of wholeness is emphasized in several different ways in Kant's later treatment of history in *The Contest of the Faculties* (1798), when Kant inquires whether it is possible to prove that there exists "a *tendency* within the human race as a *whole*" (*The Contest of the Faculties*, excerpted in *Kant: Political Writings*, here 181). Kant declares in that work that universal history is a matter of representing humanity itself as a whole, for it concerns "not only the events which may happen within a particular nation, but also their repercussions upon all the nations of the earth" (185). In *Contest of the Faculties* what is in question is humankind as an entirety: "[W]e are not dealing with any *specific* conception of mankind (*singulorum*), but with the whole of humanity (*universorum*) of humanity" (177).

[2] Kant, "Conjectures on the Beginning of Human History," in *Kant: Political Writings*, 222. Nonetheless, conjecture is, for Kant, an ambiguous procedure. He describes it as a "healthy mental recreation" for the imagination rather than "a serious activity" (221). It is required in order to fill historical gaps, but should not be whipped up into "wild conjecture [*in Muthmaßungen schwärmen*]" (222).

[3] Kant, "Conjectures," 221.

The hesitation on display here, between allowing for historical conjecture made "on the wings of imagination" and recoiling before the specter of fiction and the novel, offers a way into this chapter's focus on the formal dilemmas that beset Kant's philosophy of universal history.[4] The chapter will argue that at the cusp of modern historiography, the exigencies of Kant's universal history evoke explicitly and quintessentially aesthetic predicaments. It will begin with the observation that Kant's formal prescriptions for organizing the inchoate manifold of historical data into a narrative can be understood in light of his later *Critique of the Power of Judgment* (1790). In particular the regulative idea of progress illustrated in the "Conjectures" serves a function analogous to that of the idea of infinity in the mathematical sublime. Likewise the guiding thread of a "plan of nature" in the "Idea of Universal History" essay presages the teleological judgment of nature. In each case the demand for a guiding thread recalls the issues raised in previous chapters surrounding the representation of events as connected. Kant's formal conceit of the guiding thread [*Leitfaden*], however, gives rise to a separate conundrum: how is the representation of human freedom compatible with the formal demands of narrative continuity? For historical events must be portrayed as continuous with one another in order for the history of humanity to form a unified whole, yet freedom is defined by Kant precisely as *dis*continuous with preceding events. The spontaneity of human freedom, in other words, threatens to disrupt the coherence of the historical narrative. How is this disruptive freedom compatible with the demands of an integral narrative?

In this context, Kant's fleeting references to the novel reflect the Leibnizian provenance of universal history. What is more, they indicate that the Leibnizian thought of possible *worlds* is transformed by Kant into a thought of possible *histories*, providing a connection to Walter Benjamin's "messianic" history, one that accomplishes a leap into another future in the breach of historiographical continuity. From this, we shall see, it becomes clearer why Benjamin's disruptive

---

4 See Manfred Kuehn, *Reason as a Species Characteristic* (Cambridge: Cambridge University Press, 2009), 68, for an argument that Kant's universal history essay is about *Geschichtsschreibung*, rather than the philosophy of history per se. Timothy Bahti provides an excellent account of the background to the project of universal history and also Schiller's uptake of Kant's universal history (*Allegories of History: Literary Historiography after Hegel* [Baltimore: The Johns Hopkins University Press, 1992], 24–38). See also Albrecht Koschorke, "Codes und Narrative: Überlegungen zur Poetik der funktionalen Differenzierung," *Grenzen der Germanistik: Rephilologisierung oder Erweiterung?*, ed. Walter Erhart, 174–185 (Stuttgart: Metzler, 2004). Heinz Dieter Kittsteiner, *Out of Control: Über die Unverfügbarkeit des historischen Prozesses* (Berlin: Philo, 2004), 33, 55, refers to the *"nicht-Verfügbarkeit der Geschichte,"* i.e. the dilemma of representing history at all in a systematic fashion rather than as a mere aggregate of events.

materialist historiography, by way of Leibniz and Marburg neo-Kantianism, finds its resources in Kant's ethics, rather than in his philosophy of history. Leibniz's work on infinitesimals provides a background to the representation of freedom in Kant and also to the Marburg neo-Kantians who are significant in the context of Benjamin. This lineage provides an unexpected Kantian conduit to Benjamin's enigmatic suggestion that every second may be "the small gateway in time through which the Messiah might enter."[5]

Judgment and history

The question of how to represent human history as a coherent whole prefigures some of the difficulties that Kant later deals with in his *Critique of the Power of Judgment*. In that work Kant investigates the conditions under which we are able to "cognize" *(erkennen)* objects, including the operation of assigning subjects to predicates in order to form judgments. He narrates, in often convoluted fashion, how under normal circumstances a manifold of data of intuition is subsumed to a concept: the imagination and understanding together synthesize the manifold of intuited data, as they assign to that synthesized manifold a particular concept. The confused data of intuition thereby become a coherent, cognizable whole. Aesthetic judgment, in contrast, does not subsume a manifold of intuition to a concept, i.e., it does not succeed in "cognizing" an object and instead reflects the faculties of cognition in their free play.

Aesthetic judgment is therefore significant for Kant's critiques and for his architectonic beyond the question of what we find beautiful because it exposes the condition of judging in general, namely the cooperation of the imagination and understanding in producing a presentation by which an object is given to cognition in the first place.[6] Aesthetic judgment thus serves for Kant as an opportunity to isolate the formal process or activity of cognition, apart from any particular content or object. For aesthetic judgment does not actually judge an object

---

5 Walter Benjamin, "The Concept of History," in *Selected Writings*, vol. 4, eds. Howard Eiland and Michael W. Jennings (Cambridge: Harvard University Press, 2006), 397.
6 "[A]n aesthetic judgment is of a unique kind, and affords absolutely no cognition (not even a confused one) of the object; which happens only in a logical judgment; while the former, by contrast, relates the representation by which an object is given solely to the subject, and does not bring to our attention any property of the object, but only the purposive form in the determination of powers of representation that are occupied with it." See Kant, *Critique of the Power of Judgment*, ed. Paul Guyer, trans. Paul Guyer and Eric Matthews (Cambridge: Cambridge University Press, 2000), 113; Kant, *Kritik der Urteilskraft*, vol. 5, ed. Preußische Akademie der Wissenschaften (Berlin, Reimer, 1908); hereafter cited as AA 5.

per se, but rather responds to the subjective feeling of the judge when the faculties of imagination and understanding linger in their harmonizing play without achieving conceptualization. In a nonaesthetic judgment, the pleasure that accompanies the judgment of the beautiful is foreclosed; there is no free play of imagination and understanding because their cooperation is governed by a particular concept.

Here we see an analogy to what Hegel famously calls the "empty formalism" of Kant's deontological ethics. Kant's moral philosophy does not take as its goal to list what our duties are; instead it elucidates the conditions under which we come to know and do our duty. In analogous terms the aesthetic judgment illuminates the "empty" procedure of judgment without regard for a concept to be applied and a content to be defined.[7] Content is not at issue, in fact the arrival at a concept *derails* the free play of the imagination and understanding in the aesthetic judgment. It is therefore ultimately form alone that we judge beautiful, devoid of governance by a concept. Kant writes that aesthetic judgments cannot be produced by means of charms and ornaments, but only by pure form.[8] "Lines aimlessly intertwined" and "free designs" are examples of what we might judge beautiful because they present no recognizable content.[9] Their form alone stimulates the harmonization of imagination and understanding in their "presentational" capacities. The treatment of aesthetic judgment is hence doubly formal: it concerns the formal procedure of cognition and is occasioned by form alone.

---

[7] See Paul Guyer, *Kant on Freedom, Law, and Happiness* (Cambridge: Cambridge University Press, 2000); Wolfgang Bartuschat, "Kant über Grundsatz und Grundsätze in der Moral," *Jahrbuch für Recht und Ethik* 12 (2004): 283–298; Ido Geiger, "What Is the Use of the Universal Law Formula of the Categorical Imperative?" *British Journal for the History of Philosophy* 18, no. 2 (2010): 271–295; Laurent De Briey, "Le Formalisme pratique: De la morale à l'éthique," *Philosophiques* 32, no. 2 (2005): 319; John R. Silber, "Procedural Formalism in Kant's Ethics," *The Review of Metaphysics* 28, no. 2 (1974): 197–236; Rocío Zambrana, "Hegel's Hyperbolic Formalism," *Hegel Bulletin* 31, no. 1 (2010): 107–131; Jenny McMahon, "The Classical Trinity and Kant's Aesthetic Formalism," *Critical Horizons* 11, no. 3 (2010): 419–441; Barbara Herman, "Embracing Kant's Formalism," *Kantian Review* 16, no. 1 (2011): 49–66; Nick Zangwill, "Feasible Aesthetic Formalism," *Nous* 33, no. 4 (1999): 610–629; Rachel Zuckert, "The Purposiveness of Form: A Reading of Kant's Aesthetic Formalism," *Journal of the History of Philosophy* 44, no. 4 (2006): 599–622. Sally Sedgwick, "Hegel on the Empty Formalism of Kant's Categorical Imperative," in *A Companion to Hegel*, ed. Stephen Houlgate (Oxford: Blackwell, 2011): 265–280; Giovanni B. Sala, "Der Formalismus in der Ethik Kants: Überlegungen zu einer alten Kontroverse," *Freiburger Zeitschrift für Philosophie und Theologie* 52, nos. 1–2 (2005): 191–215.
[8] Kant, *Critique of the Power of Judgment*, 109. See Jacques Derrida, *The Truth in Painting* (Chicago: University of Chicago Press, 1987), for a discussion of Kant's mention of the *parergon*.
[9] Kant, *Critique of the Power of Judgment*, 93.

Kant's approaches to universal history are similarly embedded in an engagement with the procedural conditions of narration and with discovering the form such a history should take. Unlike the judgment of the beautiful, however, universal history does require a concept to accomplish its task of representing "this senseless course of human events" as a coherent whole. Kant's recourse, as we will show, to the concepts of progress and the plan of nature fulfills this aesthetic demand for wholeness. For in order to judge history at all, it must be conceptualized. The "senseless course of human events" must achieve sense, must appear as a unitary whole under the governance of a concept that can unify and connect all human events.

The treatment of the mathematical sublime in Kant's *Critique of the Power of Judgment* is helpful for understanding the dilemma of unifying all of human history and Kant's solution to that dilemma. In the case of the mathematical sublime, the imagination and understanding are incapable of synthesizing a manifold of data under the rule of a concept because the data are too vast or too overwhelming.[10] Kant explains, "the sublime [...] is to be found in a formless object insofar as limitlessness is represented in it."[11] Although it cannot take a sensory form, such limitlessness can be *thought* as a whole. In such a case, the thought of wholeness can arise by way of an idea that is not derived from imagination and understanding, but that is imposed upon those faculties by the higher faculty of reason. The mathematical sublime instantiates precisely such an instance of failure on the part of the faculties: we cannot *intuit* infinity with our senses. Nonetheless, we can *think* infinity; for Kant this is precisely such an instance in which reason imposes an idea upon imagination and understanding, which in their finitude are unable to present the boundlessness of infinity. In other words, reason fills in the thought of infinity where representation is foreclosed. The mathematical sublime offers a model for how an unbounded and formless manifold of data can be unified by way of a concept of reason, an "idea" which derives not from the world or our intuition of it, but rather is imposed by reason in order thereby to give to thought an unintuitable whole.

Reason's imposition of an idea in the case of the mathematical sublime is analogous to the formal procedure that Kant proposes for the composition of

---

10 "[The power of judgment], employed with regard to a presentation by means of which an object is given, requires the agreement of two powers of presentation: namely, the imagination (for the intuition and the combination of the manifold of intuition) and understanding (for the concept as presentation of the unity of this composition)" (Kant, *Critique of the Power of Judgment*, 167, translation modified).

11 Kant, *Critique of the Power of Judgment*, 128.

a universal history. A universal history must present human history as a unitary whole. All of its events therefore need to be represented as thoroughly connected to one another. Nonetheless, the data of human history are far too vast and varied to be intuited as a whole. How can such a formless, unbounded range of data as the entirety of human history be represented as forming a coherent unity? The history of humankind, as Kant indicates, cannot be *intuited* as a whole but it can be *thought* as one, if a concept of reason can be imposed to render coherent a daunting mass of unwieldy data. Hence Kant's hope in the opening of the essay on the idea of universal history that we might "discover a rule-governed progression [*einen regelmäßigen Gang*] among freely willed actions."[12] We will come back to the conflict between human freedom and history's rule later on in this chapter, but for now we simply observe the foretoken of the mathematical sublime in Kant's proposals for how to portray the course of human events.

Teleology and history

The idea of mathematical infinity – a concept of reason that is not grounded in intuition – offers the thought, or concept, of a totality to which we have no intuitive access. The idea that lets the wholeness of universal history be thought in Kant's "Idea for a Universal History with a Cosmpolitan Purpose," a concept that comes from reason rather than from understanding, is the plan of nature.[13] This idea fulfills the formal condition of representing the history of humankind, unifying it for the purposes of a historical narrative, presenting human history as a rule-governed whole.[14] It gives to conceptualization the formless data of history, just as the idea of infinity gives mathematical boundlessness to thought.

While Kant's "Idea for a Universal History" relies on the concept of the plan of nature, his "Conjectures" essay and his *Contest of the Faculties* further insist that the plan of nature must be understood as unfolding in the course of

---

12 Kant, "Idea for a Universal History," 41, translation modified.
13 This is not Frank R. Ankersmit's concept of "sublime historical experience," which has to do instead with the experience of history per se rather than with its apprehension as a whole. See Frank R. Ankersmit, *Sublime Historical Experience* (Stanford: Stanford University Press, 2005), and Ewa Domanska, "Frank Ankersmit: From Narrative to Experience," *Rethinking History* 13, no. 2 (2009):175–195.
14 See Rudolf Makkreel, *Imagination and Interpretation in Kant: The Hermeneutical Import of the Critique of Judgment* (Chicago: University of Chicago Press, 1990), 130–153.

human progress. Both the plan of nature and the idea of progress can be seen as variants on the teleology of nature analyzed in the second part of the *Critique of the Power of Judgment*. There Kant elucidates how nature, in its complexity and heterogeneity, can be conceived as a coherent whole insofar as it appears purposive to us, i.e., as organized toward a purpose or end. In that teleological judgment of nature, the manifold of data that are subsumed under the concept of nature appears as a unity, because the data are represented as governed by a purpose. The teleology of nature – the idea that nature is purposive – offers a way to see nature as a unified whole even in its infinity and in its existence over time.

The emphasis on a plan of nature and human progress serves in similar fashion to unify the narrative of the entirety of human events. In contrast to the concept of mathematical infinity, the concept of progress as the *telos* of nature offers a way to unify a manifold of *temporalized* data, i.e., to connect a sequence of events. The very concept of a plan incorporates a notion of past and future into the present and also takes a projected endpoint as the basis of the idea that unifies the manifold of nature's temporal limitlessness.[15] The plan of nature and the concept of progress thus offer formal solutions to the dilemmas of representing the history of humankind in a unified way, such that earlier and later events are connected to one another not "interstitially" but "from outside," as it were, i.e., by a governing rule. The representation of history according to the ideas of a plan of nature and progress grants a formal unity to the full range of human historical phenomena with reference to their future.[16] The reliance on these ideas in order to produce a formal unity means that at its center, universal history requires an "as if"; analogously to the teleology of nature, we regard history in accord with the idea of progress *in order to* see human history as a whole.

The idea of progress according to an "as-if" plan of nature, an idea which comes from reason rather than from understanding or indeed, from the world itself, constitutes for Kant the guiding thread by means of which to present human history as a unitary whole. Kant's formulations in the universal history essay even explicitly connect the history of humankind to the teleology of nature later highlighted in the *Critique of the Power of Judgment*, for they represent the development of the natural human faculty of rationality into the means by

---

[15] Relevant here is the treatment of "anticipated afterness" in Gerhard Richter, *Afterness: Figures of Following in Modern Thought and Aesthetics* (New York: Columbia University Press, 2011), 155. See also Kittsteiner, *Out of Control*, 150–151.

[16] I am here in complete agreement with Katerina Deligiorgi, "The Role of the 'Plan of Nature' in Kant's Account of History from a Philosophical Perspective," *British Journal for the History of Philosophy* 14, no. 3 (2006): 451–468.

which human beings make their own history. The "plan of nature" involves, on the model of Rousseau, natural human qualities that propel human beings out of the state of nature. The idea of a progress of history, embedded in a plan of nature, toward human civil union provides an ambiguous tie to the idea of a natural teleology that will come, in the *Critique of the Power of Judgment*, to represent the knowability of nature as such.

The effectivity of historical form

As we have seen, the concepts of progress and of a highest purpose of nature grant wholeness to a narrative of universal history in the same way that in the *Critique of the Power of Judgment* the concept of a teleology of nature allows us to recognize as unified "the multiplicity of the genera of earthly species and their external relations to one another."[17] Hence Kant writes in "On the Idea for a Universal History," "[T]his idea [of a plan of nature] may yet serve as a guide to us in representing an otherwise planless *aggregate* of human actions as conforming, at least when considered as a whole, to a *system*."[18] The very *idea*, in other words, of a plan of nature that results in human progress permits history to be represented as a whole whose elements cohere. The formal prescriptions for the narrative of universal history are also, however, tethered to a question of the *effectivity* of such a narrative. For Kant argues that the project of universal history would not only depict but would also *contribute to* the highest achievement of humankind, which Kant defines as "the perfect civil union of mankind," a "perfect civil constitution."[19] Historiography, Kant claims, must be regarded as "capable of *furthering* the purpose of nature itself."[20] The represented sequence of achievements of humankind should depict history and nature as having a *telos*. As a result, the narrative conceit of the intentionality of nature would promote the development of human freedom and happiness.

Here Kant breaks with the questions of narrative tactics per se and turns to the effectivity of such narratives in the world. If we can interpret human history as a development out of nature according to a plan, the representation of that history can itself sustain and contribute to our own rational moral interest. Kant's suggestion is that a unifying narrative strategy, according to the idea of a plan of nature, will recursively produce effects upon the future course of

---

17 Kant, "Idea for a Universal History," 45.
18 Kant, "Idea for a Universal History," 52.
19 Kant, "Idea for a Universal History," 51, 47.
20 Kant, "Idea for a Universal History," 51, emphasis mine.

history. The mere representation of human history as both grounded in nature and guided by nature's schemes – most notably "unsocial sociability" – actually contributes to the accomplishment of that progress: "The history of the human race as a whole *can be regarded* as the realization of a hidden plan of nature to bring about [...] [a] perfect political constitution."[21] The narrative conceit produces direct effects upon the future whose past it narrates. The presentation of human history as proceeding according to the plan of nature and directed toward a perfect civil union actually steers it in that direction. It thereby potentially *furthers* the same end of history that it *describes* as belonging to the plan of nature.

Kant implies that such a historical account would be valuable regardless of whether or not it is accurate: "[T]his idea [that nature has a "purposeful end"] might [...] prove useful."[22] Indeed, even if we were wrong about nature's purpose, it would nonetheless be *beneficial* to write that history, for "if we assume a plan of nature we have grounds for greater hopes."[23] In other words, Kant declares that the portrayal of history as unfolding toward nature's highest purpose, i.e., as human progress toward a perfect civil union, intervenes in the history it describes, actualizing the progress toward that end.[24] We can see here a precursor to Gadamer's concept of *Wirkungsgeschichte*, which historicizes the hermeneutic operation and declares the historical effectivity of texts, such that "in all understanding, whether we are expressly aware of it or not, the power of this effective-history is at work."[25] For these reasons, Max Pensky calls Kant's universal history performative and prophetic, and Barbara Herman names this a "proleptic effect," i.e., "where there is an end 'the bringing about of which is promoted by the very idea of it'."[26] This "proleptic effect"

---

21 Kant, "Idea for a Universal History," 50, emphasis mine.
22 Kant, "Idea for a Universal History," 52.
23 Kant, "Idea for a Universal History," 52.
24 Hence Pauline Kleingeld, "Kant on Historiography and the Use of Regulative Ideas," *Studies in History and Philosophy of Science* 39, no. 4 (2008): 524, asserts that "[Kant] emphasizes the productive and creative role of the theorist."
25 Hans-Georg Gadamer, *Truth and Method*, trans. Joel Weinsheimer and Donald G. Marshall (New York: Continuum, 2012), 300. See also Sinéad Murphy, *Effective History: On Critical Practice under Historical Conditions* (Evanston: Northwestern University Press, 2010); Csaba Olay, "Die Überlieferung der Gegenwart und die Gegenwart der Überlieferung: Heidegger und Gadamer über Tradition," *International Yearbook for Hermeneutics* 12 (2013): 196–219.
26 Max Pensky, "Contributions Toward a Theory of Storms: Historical Knowing and Historical Progress in Kant and Benjamin," *The Philosophical Forum* 41, no. 1–2 (2010): 159; Barbara Herman, "A Habitat for Humanity," in *Kant's Idea for a Universal History with a Cosmopolitan Aim: A Critical Guide*, eds. James Schmidt and Amélie Rorty (Cambridge: Cambridge University

means that "seeing the social world as tending toward a final end is essential to making it true that it reaches it."[27]

Freedom, spontaneity, and the infinitesimal

Kant's declaration that the writing of universal history would intervene in the course of history recalls Foucault's characterization of Kant's "What is Enlightenment?" as an intervention in modernity, "an experiment with the possibility of going beyond [the limits that are imposed upon us]," that "patient labor giving form to our impatience for liberty."[28] Foucault sees Kant's "What is Enlightenment?" in a performative light, evoking new possibilities for freedom. Nonetheless, it is difficult to reconcile Kant's concern for human freedom with his references to the plan of nature.[29] The one-by-one connection of events comes into tension with the overall unity that is accomplished by the ideas of progress and of a plan of nature. For by Kant's definition, freedom introduces unpredictability into our perspective: "We are not able to adopt an absolute point of view when trying to predict free actions."[30] Freedom is for Kant a spontaneous causality; it is not determined by antecedents, unlike the causality of nature according to which a sequence of events is governed by natural laws and in which each event is the determinate outcome of preceding factors. We can trace Kant's representation of a rupture in causality at least as far back as Luther. The spontaneity of Kantian freedom resembles the description of the Christian's freedom in Luther, in which the believer performs good works

---

Press, 2009), 154; see also Dominick LaCapra, "Kant, Benjamin, Pensky and the Historical Sublime," *The Philosophical Forum* 41, nos. 1–2 (2010): 177.
27 Herman, "Habitat for Humanity," 164.
28 Michel Foucault, "What is Enlightenment?" in *The Foucault Reader*, ed. Paul Rabinow (New York: Pantheon Books, 1984), 50.
29 Peter Gilgen, *Lektüren der Erinnerung: Lessing, Hegel, Kant* (Munich: Fink, 2012), 108–109, argues that the autonomy of moral action interferes with the systematic representation of universal history. Paul Guyer argues that they are, for Kant, ultimately integrated within Kant's system, see Guyer, "Ends of Reason and Ends of Nature: The Place of Teleology in Kant's Ethics," in *Kant's System of Nature and Freedom: Selected Essays* (Oxford: Oxford University Press, 2005), 169–197; Jennifer Mensch, *Kant's Organicism: Epigenesis and the Development of Critical Philosophy* (Chicago: University of Chicago Press, 2013), 139, explains that the concept of epigenesis provides a decisive link between Kant's claims for the teleology of nature and the development of human rationality and freedom, so that reason could be characterized as "self-born."
30 Kant, *Contest of the Faculties*, excerpted in *Kant: Political Writings*, ed. Hans Reiss, trans. H.B. Nisbet (Cambridge: Cambridge University Press, 1992), 180.

"out of spontaneous love," but a spontaneous love that is the product of "obedience to God," where such obedience is itself a gift of grace, the result of God's own choice to "write his law in our hearts."[31] In each case, spontaneity is the paradoxical result of submission: for Luther spontaneous love produces good works resulting from a submission to the will of God; for Kant, moral acts result from the will's submission to the governance of pure practical reason and hence to the possibilities of freedom.[32]

To be clear, human freedom is for Kant grounded in reason, where reason is a cause, but specifically a cause independent of any preceding event.[33] Kant's teleology of progress would therefore seem to speak against the proximal autarkic causality of freedom.[34] The appearance of punctual freedom in the

---

[31] Luther, "The Freedom of a Christian," in *Martin Luther's Basic Theological Writings*, ed. Timothy F. Lull (Minneapolis: Fortress Press, 1989), 611, 628.

[32] On Kant and Luther see Bernard Wand, "Religious Concepts and Moral Theory: Luther and Kant," *Journal of the History of Philosophy* 9, no. 3 (1971): 329–348; Jörg Wurzer, "Der Freiheitsbegriff Martin Luthers und Immanuel Kants im Vergleich," *Luther: Zeitschrift der Luther-Gesellschaft* 71, no. 1 (2000): 21–35.

[33] Kant, *Critique of Pure Reason*, ed. and trans. Paul Guyer and Allen W. Wood (Cambridge: Cambridge University Press, 1998), 540 [A547/B575]. In his early work Kant portrays free will in a less interruptive, autarkic fashion. He embraces a Leibnizian formulation in his *Nova Dilucidatio* (1755), arguing that the freedom of the will does not amount to a freedom from causal determination. Rather, it consists in the will being able to choose whatever seems best, a "rational spontaneity" that is constrained by the good; see David Forman, "*Appetimus Sub Ratione Boni*: Kant's Practical Principles between Crusius and Leibniz," in *Kant und die Philosophie in Weltbürgerlicher Absicht* eds. Stefano Bacin et al. (Berlin: De Gruyter, 2015), 325. This rational spontaneity stands in contrast to an absolute spontaneity that would be in principle arbitrary, i.e. governed by no rule whatsoever. Later in his career Kant comes to sympathize with Crusius's rejection of Leibniz and Wolff's compatibilism (Forman, "*Appetimus Sub Ratione Boni*," 324) and advocates an "absolute freedom" in which the will is inclined, but not determined, by the good. Under this model freedom is understood as discontinuous with the chain of natural causality and also not fully determined by rational causality, i.e., by the absolute constraint of the good. The topic of spontaneity is also central in Kant's account of the receptivity vs. generativity of the mind. See Robert B. Pippin, "Kant on the Spontaneity of Mind," in *Hegelian Variations: Idealism as Modernism* (Cambridge: Cambridge University Press, 1997), 29–55; David Allison, "Spontaneity and Autonomy in Kant's Conception of the Self," in *The Modern Subject: Conceptions of the Self in Classical German Philosophy*, eds. Karl Ameriks and Dieter Sturma (Albany: SUNY Press, 1995), 11–30; Marco Sgarbi, *Kant on Spontaneity* (London: Continuum, 2012), 61–78; Melissa McBay Merritt, "Reflection, Enlightenment, and the Significance of Spontaneity in Kant," *British Journal for the History of Philosophy* 17.5 (2009): 981–1010. Georg Wallwitz, "Kant über Fatalismus und Spontaneität," *Allgemeine Zeitschrift für Philosophie* 28, no. 3 (2003): 207–228; Dirk Setton, "Absolute Spontaneity of Choice," *Symposium* 17, no. 1 (2013): 75–99.

[34] Emil Fackenheim, *Metaphysics and Historicity* (Milwaukee: Marquette University Press, 1961), 40, remarks that Kant's own concern is for the "form of a quasi-history," insofar as

course of human history would likewise seem to form an obstacle to representing the temporal sequence of events as coherent and connected. For if human history is to be perceived as proceeding according to the rule-governed plan of nature, then how could spontaneity appear punctually within it? If freedom is an autarkic causality and thus without alien determinations, its narrative would need to be *dis*continuous, for by Kant's definition *no* causality connects one free act to another nor to any preceding event at all. Freedom poses a formal problem to the narrative of universal history, threatening its continuity and its unity. Genuine freedom would seem to defy the requirement of a cohesive form when it comes to Kant's quest for a rule-governed representation of human history held together by means of a predetermined concept.

The disruptive discontinuity of freedom is thematized more overtly in the *Contest of the Faculties*. There Kant recognizes the tension between the freedom of the individual and the progress of humankind as a whole, in view of the question whether predictions for the future are possible. Responding to the question "Is the human race constantly progressing?" Kant emphasizes the unpredictability of free actions, acknowledging the conflict between representing autarkic human action as unpredictable and representing the course of history as continuous and coherent: "[W]hile [man] needs to perceive a connection governed by natural laws before he can foresee anything, he must do without such hints or guidance when dealing with *free* actions in the future."[35] Kant comes to an unsatisfying resolution: when we as spectators of human history exhibit enthusiasm for developments toward a just civil constitution, this provides indirect evidence of our morality and thus confirms that human history can be said to be guided by reason.[36] This does not, however, mitigate Kant's acknowledgment that autarkic freedom is as potentially disruptive of historical continuity as it is fundamental to morality.[37]

"such a form is required by the doctrine [...][of] the appearance, and development, of freedom." "Form" and "quasi-history" are here at odds insofar as freedom marks discontinuity and form marks cohesion.
35 Kant, *Contest of the Faculties*, 181, emphasis is Kant's. The emergence of modern statistics is a significant factor here, for statistical probability makes predictibility possible and threatens to infringe on the unpredictability of human action. See David Martyn, "Figures of the Mean: Freedom, Progress, and the Law of Statistical Averages in Kleist's 'Allerneuester Erziehungsplan'," *The Germanic Review: Literature, Culture, Theory* 85, no. 1 (2010): 44–62.
36 Kant, *Contest of the Faculties*, 182.
37 See John H. Zammito, "A Text of Two Titles: Kant's 'A Renewed Attempt to Answer the Question: "Is the Human Race Continually Improving?"'" *Studies in History and Philosophy of Science, Part A* 39, no. 4 (2008): 535–545.

Continuity and the novel of history

Kant's philosophy of universal history rests on the unifying principle of a plan of nature. At the same time, freedom troubles the continuity of the narrative. How therefore can freedom be accounted for in any continuous narrative form, when it functions as a *dis*connection from any foregoing events and causalities? Let us briefly consider how these narrative predicaments are rooted in Leibnizian metaphysics. Although continuity was already a problem for ancient Atomism, the provenance of Kant's dilemmas surrounding causal connectibility can be seen in Leibniz's rejection of occasionalism. This doctrine, articulated most clearly by Nicolas Malebranche in the seventeenth century, held that God is the sole and true cause or agent that mediates between what are intrinsically disconnected events.[38] Leibniz's doctrine of monads, in a similar fashion, famously denies genuine interaction between created substances. This metaphysical discontinuity, in which substances are presumed to be entirely unaffected by one another, preserves an occasionalist model of a causal disconnection of events. Nonetheless, Leibniz's notion of a predetermined harmony compensates for that Malebranchian absolute discontinuity by subordinating all creation to God's will.[39] Leibniz's claims for the independence of monads, which nonetheless are each rooted in a predetermined harmony of the universe, offers a model for Kant's postulation of individual freedom within a course of human events that is guided by a plan of nature.

As we have seen in the previous chapters, the tension between unity and discontinuity exposes the proximity of literary and historiographical issues. Both Leibniz and Kant offer passing reflections on the significance of literature to metaphysical questions. Kant toys fleetingly with the connection between the novel and historiography, writing,

> Es ist zwar ein befremdlicher [...] Anschlag, nach einer Idee, wie der Weltlauf gehen müßte, wenn er gewissen vernünftigen Zwecken angemessen sein sollte, eine Geschichte abfassen zu wollen; es scheint, in einer solchen Absicht könnte nur ein Roman zu Stande kommen.

---

[38] See Dominik Perler and Ulrich Rudolph, *Occasionalismus: Theorien der Kausalität im arabisch-islamischen und im europäischen Denken* (Göttingen, Germany: Vandenhoeck & Ruprecht, 2000); Lisa Downing, "Occasionalism and Strict Mechanism: Malebranche, Berkeley, Fontenelle," in *Early Modern Philosophy: Mind, Matter, and Metaphysics*, eds. Christia Mercer and Eileen O'Neill, 206–230 (Oxford: Oxford University Press, 2005); and Steven M. Nadler. "Occasionalism and General Will in Malebranche," *Journal of the History of Philosophy* 31, no. 1 (1993): 31–47.

[39] For a treatment of the relationships to literature of Leibniz and Malebranche, see Frédéric de Buzon, "Littérature et Fiction: Leibniz et Malebranche," *Dix-septième Siècle* 255, no. 2 (2012): 241–256.

It is admittedly a strange [...] proposition to write a history according to an idea of how world events must develop if they are to conform to certain rational ends; it would seem that only a novel [ein Roman] could result from such premises.[40]

We have seen in this chapter's earlier discussion of the role of conjecture that for Kant a history based "solely on conjectures would seem little better than drawing up a plan for a novel," and the resulting work would constitute "a work of fiction."[41] These references, however, do not amount to a thoroughgoing rejection of literary and novelistic elements. On the contrary, Seán Williams has shown in detail that Kant's philosophical history involves an "overarching *narrative* argument" while at the same time Kant's gesture of rejection of the novel constitutes a "witty *topos* of Kant's mature thought."[42] In other words, a narrative of universal history argues, *in its very narrative form*, for the existence of progress. The narrative form itself *is* an argument of a sort.

Kant's mention of the novel in this context provides another link to Leibnizian metaphysics.[43] In the *Theodicy* (1710) Leibniz remarks that the history of humanity is a novel in the mind of God.[44] It is worth noting that this remark appears in a context that pertains to the problem of continuity within a whole. At issue is Pierre Bayle's assertion that there are good and bad people and principles: Leibniz argues against Bayle, to the effect that the history of

---

40 Kant, "Idee zu einer allgemeinen Geschichte im weltbürgerlicher Absicht," in AA 8:29; "Idea for a Universal History," 51–52.
41 Kant, "Conjectures," 221. As Kuehn argues, Kant is not offering a mere piece of fiction nor a novel (Kuehn, "Reason as a Species Characteristic," 69).
42 Seán M. Williams, "Kant's Novel Interpretation of History," *Seminar: A Journal of Germanic Studies* 49, no. 2 (2013): 171–90, 171 (emphasis mine), 183; see also Chad Wellmon, *Becoming Human: Romantic Anthropology and the Embodiment of Freedom* (University Park, PA: Pennsylvania State University Press, 2010), 167.
43 Leibniz's theory of possible worlds is of course significant for the emergence of the modern novel. See Sébastien Charles, "Le possible comme critique du Spinozisme: Leibniz et la fiction," *Science et Esprit: Revue de Philosophie et de Théologie* 67, no. 1 (2015): 17–33. Rüdiger Campe, *The Game of Probability: Literature and Calculation from Pascal to Kleist*, trans. Ellwood H. Wiggins, Jr. (Stanford: Stanford University Press, 2012), 248–272; Eugene F. Kaelin, "Language as a Medium for Art," *Journal of Aesthetics and Art Criticism* 40 (1981): 121–130; Sylviane Malinowski-Charles, "De la possibilité des fictions littéraires chez Spinoza," *Teoria: Rivista di Filosofia* 32, no. 2 (2012): 247–265.
44 G.W. Leibniz, *Theodicy: Essays on the Goodness of God, the Freedom of Man, and the Origin of Evil*, ed. Austin Farrer, trans. E.M. Huggard (La Salle, IL: Open Court, 1985), 217. See Reinhart Koselleck, "Historia Magistra Vitae: The Dissolution of the Topos into the Perspective of a Modernized Historical Process," in *Futures Past: On the Semantics of Historical Time*, trans. Keith Tribe (Cambridge: MIT Press, 1985), 30. Leibniz's theory of possible worlds is of course significant for the emergence of the modern novel.

humanity as a whole is good. His belief about the good of the whole requires Leibniz to reject the notion that within human history there is a discontinuity resulting from two different principles that would govern two distinct elements. What Bayle perceives as an inconsistent distribution of bad and good events in human history, Leibniz reframes as a consistent, continuous, and good whole. The image of the novel in the mind of God thus grants history not only the quality of goodness, but also the quality of unity.[45] The image resolves as a matter of perspective the tension between the discontinuous spontaneity of human freedom and the continuity of a history governed by the idea of progress, which explains its significance to Kant:

> Jenes hat daher Ursache, alle Übel, die es erduldet, und alles Böse, das es verübt, seiner eigenen Schuld zuzuschreiben, zugleich aber auch als ein Glied des Ganzen (einer Gattung) die Weisheit und Zweckmäßigkeit der Anordnung zu bewundern und zu preisen. (AA 8:116)
>
> The individual therefore has cause to blame himself for all the ills which he endures and for all the evil which he perpetrates; but at the same time, as a member of the whole (of a species), he has cause to admire and praise the wisdom and purposiveness of the overall arrangement. (Kant, "Conjectures," 227; *sic* parenthesis)

For Kant, we are the sources of genuine evil, as in Luther's rendition, but history can nonetheless be viewed as a cohesive whole insofar as we form part of the whole of our species.

Humboldt's mediated representation

Wilhelm von Humboldt offers an example of how the Kantian formal concerns for providing connections and a holistic narrative appear in nineteenth-century historicism, for Humboldt exhibits a thoroughgoing formalism in his historical thought.[46] Humboldt begins his 1821 lecture on the task of the historian by defining that task as the "representation of what happened [*Darstellung des Geschehenen*]," and in this respect Humboldt also appears as a pinnacle of the kind of positivist historicism that Benjamin associates with Ranke, as we will see in the next chapter. But Humboldt goes on to explain just how mediated

---

45 Heinz Dieter Kittsteiner, "Walter Benjamin's Historicism," *New German Critique* 39 (1986): 201n70, alludes to the idea of history as "God's hieroglyph" in response to history as a 'cipher that is to be read', an idea from Herder that remained prominent through early Romanticism.
46 White, *Metahistory: The Historical Imagination in Nineteenth-Century Europe* (Baltimore: The Johns Hopkins University Press, 1973), 184.

and aesthetic that representation is, insofar as it must achieve wholeness and bridge lacunae. For after the pronouncement that history is the representation of what happened, Humboldt writes that "an event [...] is only partially visible in the world of the senses[,] the rest has to be added by intuition, inference and guesswork."[47] The references to intuition, inference, and guesswork invoke techniques of imagination to fill gaps, i.e., to synthesize a whole that would reflect in a mediated fashion how events truly happened.[48]

In emphasizing that the fragments of history must be collected and imaginatively put into a shape of a whole, Humboldt is, surprisingly, closer to the *Critique of the Power of Judgment* than to Kant's essay on universal history. For Humboldt, however, freedom is not the central issue that it is for Kant and the course of history is not a matter of subjective purposiveness but instead of laws. That is, for Humboldt necessity signifies unchangeable mechanical laws that determine the sequence of events. Hence Humboldt declares that the historian's job is to formulate an idea that explains the necessary connection between what may appear to be a mere "casual concatenation of events."[49] This notion of laws that determine the connections between events in the historical narrative illuminates Karl Popper's claims for the totalitarian aspects of Humboldt and historicist thinking and, as we will see in Chapter Seven, Arendt's criticism of the logic and metaphorics of "process" in conceiving of history. If history is governed by laws, our human freedom is thereby nullified. Nonetheless, Humboldt allows, with his reference to guesswork in particular, an "as if" qualification for the truth of the historian, which accords with the "as if" quality of Kant's critical philosophy made famous by Hans Vaihinger's 1911 *The Philosophy of "As If."*[50]

Indeed, Humboldt pays careful attention to aesthetic and literary attributes. He attests that imagination and qualities similar to the poet's are required to produce history, since language is not free of connotations and so narratives are not "literally true."[51] He likewise asserts that "the activities of

---

47 Wilhelm von Humboldt, "On the Historian's Task," *History and Theory* 6, no. 1 (1967): 57.
48 Koselleck, *Futures Past*, 30–31, describes how Humboldt dissolved the centuries-old conflict between poetics and history, suggesting that there must be a "history in general" which Koselleck considers "a criterion of epic representation" that Humboldt transforms into a category of the historical.
49 Claus Uhlig, "Literature as Textual Palingenesis: On Some Principles of Literary History," *New Literary History* 16, no. 3 (1985): 490.
50 Hans Vaihinger, *Die Philosophie des Als Ob: System der theoretischen, praktischen und religiösen Fiktionen der Menschheit auf Grund eines idealistischen Positivismus* (Leipzig: Meiner, 1922).
51 Humboldt, "Historian's Task," 58.

[the historian and the poet] are undeniably related."[52] Exact knowledge is necessary for the historian, but as Paul R. Sweet has explained in an article on Humboldt, passion, native talent, and something beyond what Humboldt calls "cold prosaic thought," namely an "intuitive reach," are also necessary.[53] These creative and intuitive elements engender, in Humboldt's description, not only a narrative but a form of effectivity. He writes that the language of the historical narrative would provide "a unique vehicle for the creation and communication of ideas."[54] As with Kant's claims for the effects the mere production of a narrative of universal history could have, Humboldt suggests that in this form of history language does not merely represent but rather *creates* and inaugurates ideas in the public sphere.[55] Humboldt is here strangely close to Kant – both the Kant of the third critique and of the "Idea for a Universal History" – despite his ties to positivist historicism.

Benjamin and Kant

So far this chapter has examined the intersection of Kant's philosophy of history and the operations of judgment, in particular pertaining to the mathematical sublime and the teleology of nature. It has argued that there is a tension between the demands of narrative continuity and the disruptive character of freedom, which is illuminated with reference to Leibniz's doctrine of monads and predetermined harmony. Let us observe briefly the legacy of these various strands in Benjamin, whom we will discuss at greater length in the next chapter in the context of Ranke. The genealogy from Leibniz through Kant to Benjamin is thoroughly mediated by Marburg neo-Kantians, among whom continuity was a significant issue. Paula Schwebel has shown in detail the extent to which Benjamin was influenced by radically divergent receptions of Leibniz and the doctrine of monads by Hermann Cohen and Heinz Heimsoeth.[56] John H. Smith has elucidated the significance of the infinitesimal from Leibniz and Kant

---

52 Humboldt, "Historian's Task," 58.
53 Paul R. Sweet, "Wilhelm Von Humboldt (1767–1835): His Legacy to the Historian," *The Centennial Review* 15, no. 1 (1971): 31, 35.
54 Humboldt, "Historian's Task," 70. Humboldt also betrays some idealist moments; he writes that there is an "inexorable" process of history and that "[t]he goal of history can only be the actualization of the idea which is to be realized by mankind."
55 This is one of the main claims of Sweet, "Wilhelm Von Humboldt."
56 Paula L. Schwebel, "Intensive Infinity: Walter Benjamin's Reception of Leibniz and its Sources," *MLN* 127, no. 3 (2012): 589–610.

through Ernst Cassirer and Cohen, demonstrating that neo-Kantianism ultimately recast continuity as a quasi-Kantian principle of cognition, as demonstrated in these lines of Cohen: "Continuity is a law of thought. It is the law of thought of the connection which enables the generation of the unity of knowledge and thereby the unity of the object of knowledge."[57]

Benjamin's relationship to neo-Kantianism and Kant is nonetheless not straightforward. In a May 1918 letter to the composer Ernst Schoen, Benjamin writes that he studies Kant because he is among the "greatest adversaries" of revolutionary thoughts, a "despot," who has "philosophized certain insights that are among the reprehensible ones to be found in ethics in particular."[58] Benjamin's "Program for a Coming Philosophy" (1918) gives some clues to what, as he writes to Schoen, makes Kant a "despot," above all his projects of justifying [*rechtfertigen*] knowledge and grounding reason rather than working through their relationship to experience.[59] Thus Benjamin writes that Kant that "drives and senselessly whips his hobbyhorse, the logos."[60] In his 1965 lecture course on freedom and history, however, Adorno depicts Benjamin as ambivalent toward Kant. He defines as one of the "underlying motifs" in Benjamin's thought an "attempt to differentiate himself from Kant [...] who impressed him and appeared very powerful, but was also, I should like to add, something of a threat."[61] Benjamin also writes to Schoen that "Kant's prose per se represents

---

[57] Quoted in Thomas Mormann and Mikhail Katz, "Infinitesimals as an Issue of Neo-Kantian Philosophy of Science," *HOPOS: The Journal of the International Society for the History of Philosophy of Science* 3, no. 2 (2013): 266. John H. Smith, "Kant, Calculus, Consciousness, and the Mathematical Infinite in Us," *Goethe Yearbook* 23 (2016): 116n8, argues that Kant represents a central move in modernity to tame the dangerous idea of infinity. See also Smith, "The Infinitesimal as Theological Principle: Representing the Paradoxes of God and Nothing in Cohen, Rosenzweig, Scholem, and Barth," *MLA* 127, no. 3 (2012): 562–588; see also Pierfrancesco Fiorato, "'Zeitlos und dennoch nicht ohne historischen Belang': Über die idealen Zusammenhänge der Geschichte bei dem jungen Benjamin und Hermann Cohen," *MLN* 127, no. 3 (2012): 611–624; Julia Ng, "Walter Benjamin's and Gershom Scholem's Reading Group around Hermann Cohen's *Kants Theorie der Erfahrung* in 1918: An Introduction." *MLN* 127, no. 3 (2012): 433–439; Peter Eli Gordon, "Science, Finitude, and Infinity: Neo-Kantianism and the Birth of Existentialism," *Jewish Social Studies* 6, no. 1 (1999): 30–53.
[58] Walter Benjamin, *The Correspondence of Walter Benjamin, 1910–1940*, eds. Gershom Scholem and Theodor Adorno, trans. Manfred R. Jacobson and Evelyn M. Jacobson (Chicago: University of Chicago Press, 1994), 125.
[59] Benjamin, *Correspondence*, 157.
[60] Benjamin, *Correspondence*, 125; see also Howard Caygill, *Walter Benjamin: The Colour of Experience* (London: Routledge, 1998), 23.
[61] Theodor Adorno, *History and Freedom: Lectures 1964–1965*, ed. Rolf Tiedemann, trans. Rodney Livingstone (Cambridge: Polity, 2008), 145. Adorno points out that later Benjamin

the limes of literary prose."[62] Despite his vehement response to Kant's philosophy of history, Benjamin's "Program for a Coming Philosophy" can be read as "supplementing" Kant's transcendental logic rather than rejecting it outright.[63] Hence in 1917 Benjamin writes, "no matter how great the number of Kantian minutiae that may have to fade away, Kant's system's typology must last forever [...] what is essential in Kant's thought must be preserved."[64] This ambivalence toward Kant fits the neo-Kantian context of Benjamin's time, for instance its maxim of "with Kant beyond Kant" and Paul Natorp's rejection of Kant's fundamental separation of sensibility and understanding.[65]

Kant's attempts to formulate a universal history do not, for Benjamin, belong to what is essential in his thought. Benjamin abandoned an early idea for a dissertation on Kant's philosophy of history because, as he writes to Scholem in 1917, he was disappointed in the essays "On the Idea for a Universal History with a Cosmopolitan Purpose" and "Perpetual Peace."[66] Nonetheless, Benjamin's disdain for Kant's universal history essay, and for the transcendental project of grounding knowledge and reason, does not amount to a wholesale rejection of Kant as a resource for the philosophy of history. In Kant's ethics Benjamin finds surprising resources, writing to Scholem,

> Von Kants *historischen* Schriften aus einen Zugang zur Geschichtsphilosophie zu gewinnen ist schlechterdings unmöglich. Anders wäre es von der Ethik aus; auch das ist nur beschränkt möglich und Kant selbst ist diesen Weg nicht gegangen.[67]
>
> It is virtually impossible to gain any access to the philosophy of history using Kant's *historical* writings as a point of departure. It would be different if the point of departure were his ethics; even that is possible only within limits. Kant himself did not travel this path.[68]

---

does not dispute Kant, but passes over his universal history in silence even where he critiques the Social Democrats' concept of progress and the notion of the improvement of humankind. Benjamin does discuss Kant in some detail, for instance in a 1938–1939 review, unpublished in Benjamin's lifetime, of Richard Hönigswald's *Philosophie und Sprache: Problemkritik und System*, see *Selected Writings*, vol. 4 (Cambridge: Harvard University Press, 2003), 139–144.

62 Benjamin, *Correspondence*, 98.
63 Caygill, *Colour of Experience*, 25.
64 Benjamin, *Correspondence*, 97.
65 Quoted in Mormann and Katz, "Infinitesimals," 238; 241.
66 Benjamin, *Correspondence*, 105. See also Caygill, *Colour of Experience*, 155n31.
67 Walter Benjamin, *Gesammelte Briefe*, vol. 1, eds. Theodor Adorno and Gershom Scholem (Frankfurt am Main: Suhrkamp, 1978), 176, emphasis mine.
68 Benjamin, *Correspondence*, 116, *sic* emphasis.

Why does Benjamin take an interest in Kant's ethics rather than in his overtly historical writings when it comes to the formulation of a philosophy of history?[69]

Kant's second critique investigates the pure practical reason that can guide the will and asks the question of how spontaneous willed action happens at all, i.e., action that is not predetermined by any events preceding it – which would certainly confound the historiographer's attempt to describe a "rule-governed progression." These portrayals of reason and freedom raise the problem of how to narrate a cohesive history while acknowledging the interruptive and self-determining capacity of freedom.[70] What is essential in Kant's thought for Benjamin when it comes to history is his ethical thought, because it is concerned precisely with the punctual and spontaneous intervention of human freedom, which would seem to be at odds with the formal demands of historiography. For insofar as Kant does portray human freedom as a spontaneous causality, such an interruption in the connection of events would disrupt the unity and narrative continuity that Kant demands of universal history. Thus the tension between the structure of narrative on the one hand and the contingency of actual events and human agency on the other hand troubles the requirements of narrative continuity in the project of universal history. Precisely that troubling effect is at the heart of Benjamin's materialist historiography, with its constellating, interruptive technique.

As we have seen, for Kant universal history should represent the manifold of human events as a unitary whole, under the governance of the idea of progress, in turn rooted in a plan of nature. The emphasis on continuity and purposiveness, however, evokes a predicament with respect to how freedom appears in the narrative, when freedom is conceived as a spontaneity that ruptures a continuity of events. This endemic historiographical rupture is more obviously thematized in Benjamin's materialist historiography, but can be seen to appear already in Kant's considerations of freedom, causality, the eruption of spontaneity, and the effectivity of historiography itself. Universal history raises quintessentially literary questions about the operations of narrative in service of philosophy. The furtherance of human freedom seems to call for a narrative of progress, which effectively suppresses the interruptive capacity of such freedom.

---

[69] See on this point Beatrice Hanssen, *Walter Benjamin's Other History: Of Stones, Animals, Human Beings, and Angels* (Berkeley: University of California Press, 1998), 26–30.

[70] Peter D. Fenves, *The Messianic Reduction: Walter Benjamin and the Shape of Time* (Stanford: Stanford University Press, 2011), 152–186, describes in detail the connections between Benjamin's "Program" essay, Marburg neo-Kantianism, and Husserl. Further treatment appears in Fenves, *Arresting Language: From Leibniz to Benjamin* (Stanford University Press, 2001), 205–215.

5 Not Benjamin's Ranke: On the Aesthetics of Historicism

In the previous chapter we observed the narrative implications of Kant's philosophy of universal history and its concern for continuity and structural emplotment. Leopold Ranke, a central figure of nineteenth-century German historicism, writes actual history but also formulates a philosophy of history that reacts to the progressive continuity of history portrayed by Kant and Hegel. In contemporary literary studies, Ranke is perhaps best known by way of Walter Benjamin. Benjamin's "On the Concept of History" famously quotes Ranke's already-famous line that writers of history must seek to portray "wie es eigentlich gewesen," or "how it really was."[1] The phrase "how it really was" has long been taken to epitomize the epistemological naïveté of Ranke and his ilk. Similarly, Nietzsche's characterizations of nineteenth-century historiography as "stupefying [betäubend]" and as a kind of illness [Krankheit] are crucial in the twentieth-century reception of Ranke.[2] A century after Ranke, for instance, Adorno cites Ranke's "as it really was" as the dictum of an untheorized positivist historiography, according to which any broader interpretation of history is forbidden and from which the philosophy of history is excluded.[3] "As it really was" is also seen to reflect Ranke's obtuseness with respect to the artistic task of the historian, including the formal aspects of narrative that animate historiography. In this vein, Georg Simmel writes in 1907 that nineteenth-century thinking about history is oblivious to "the presuppositions of the historical construction that lies beyond everything historical."[4]

Both the broad-scale interpretation of history and the procedures of historical construction to which Adorno and Simmel refer revolve around the question of how events come to be connected in an intelligible way within a

---

[1] Leopold von Ranke, "Preface to the First Edition of *Histories of the Latin and Germanic Peoples*," in *The Theory and Practice of History*, ed. Georg G. Iggers, trans. Wilma A. Iggers (New York: Routledge, 2011), 86, translation modified; *Geschichten der romanischen und germanischen Völker von 1494 bis 1515*, in *Sämtliche Werke*, vol. 33–34 (Leipzig: Duncker und Humblot, 1885), 7. Edition referred to hereafter as *SW*, followed by volume and page number.
[2] Friedrich Nietzsche, *Unfashionable Observations*, trans. Richard T. Gray (Stanford: Stanford University Press, 1995), 135, 163.
[3] Theodor Adorno, *History and Freedom: Lectures 1964–1965*, ed. Rolf Tiedemann, trans. Rodney Livingstone (Cambridge: Polity Press, 2006), 10–11.
[4] Quoted in Heinz Dieter Kittsteiner, "Walter Benjamin's Historicism," *New German Critique* 39 (1986), 190.

historiographical account. For this reason, Ranke's reflections on historical continuity are significant for the present work. This chapter will argue that the epistemological and aesthetic dilemmas of narrative connection are fully on display within Ranke's own thought, contrary to the claims of his detractors. For instance, although Benjamin's broad critique of historiography seems to lump Ranke together with triumphalist and progressivist models of history, in fact Ranke, in the tradition of Herder, rejects universal history, in which moments and eras are represented as part of a linear trajectory of improvement. Instead, he holds to a motto that also counts among his best known dicta, namely: "Every epoch is immediate to God [*Jede Epoche ist unmittelbar zu Gott*]."[5] This dictum opens Ranke to accusations of extreme historical relativism, but for the purposes of this chapter the dictum is significant in its insistence on the distinctness of epochs and its implicit rejection of the progressivist philosophy of history.

As we will see in the pages that follow, Ranke's purview of historiographical topics includes aesthetic, epistemological, and narrative issues that are intrinsically bound up with the question of how to represent connections between epochs and events. In some respects, his work bears traces of the universal-historical gesture, despite Ranke's insistence to the contrary, but we also observe Ranke's concern for the aesthetics and poetics of storytelling per se. For Ranke was epistemologically and aesthetically far more shrewd than his general reception would have him appear. The characterizations of Ranke by Adorno and Simmel disregard Ranke's anti-progressivist, aesthetic, formalist, disruptive, and anti-ideological gestures. With regard to a sense of form and literary skill, we will see that Ranke makes use of implicitly aesthetic principles that foreclose a naïve epistemological approach to the representation of history "as it really was." In closing, this chapter will observe the degree to which Ranke characterizes his approach to history as a rejection of older historical writing, in terms that resemble Benjamin's rejection of Ranke. For Ranke portrays his very own predecessors as inadequately imaginative and as naïve, even dangerously so, with respect to the exigencies of historical representation. He also portrays them as improperly eliding the differences between singular historical moments. Indeed, Heinz Dieter Kittsteiner has argued that Benjamin's version of historical materialism is much closer to Ranke's historicism than may initially appear.[6] In line with Kittsteiner,

---

5 Ranke, "On Progress in History (From the first lecture to King Maximilian II of Bavaria 'On the Epochs of Modern History, 1854')," in *Theory and Practice of History*, 21.
6 And conversely, according to Kittsteiner in "Walter Benjamin's Historicism," Ranke is more messianic than we might expect.

Leonard Krieger, Peter Burke, and others, this chapter will show that Ranke is "more complicated than the symbol he became."[7]

Form in Hegelian historicism

It is important to note that the term "historicism" is something of a moving target. Ranke's historicism is not to be confused, for instance, with the historicism that Karl Popper in 1957 associates with Hegel, which depicts history as following an inexorable trajectory according to universal laws.[8] Popper enumerates a litany of dangers associated with Hegel's philosophy of history. His main problem with Hegelian historicism is that it presumes that events are incontrovertibly connected.[9] Popper holds such an understanding of history responsible for twentieth-century communism and National Socialism. This criticism of Hegel, however, does not represent the range of meanings of the designation "historicism" (nor is Popper's characterization of Hegel universally accepted). The use of the English word "historicism" to translate *Historismus* eclipses the distinction between *Historismus*, historicism, and the variations within each of these, as Beiser's magisterial *The German Historicist Tradition* relates in detail.[10]

We have seen in the previous chapter that for Kant the ideas of progress and a plan of nature operate in service of the production of a unified, cohesive, and effective account of history, formed narratively according to the rational ideas of progress and the plan of nature. Hegel, in contrast, famously historicizes thought, philosophy, right, and art, portraying them as evolving within particular periods, according to particular circumstances and particular configurations of *Geist*, or spirit. Although he is identified as a German Idealist, in fact Hegel deals with how *Geist actually* unfolds dialectically and teleologically

---

7 See Leonard Krieger, "Elements of Early Historicism: Experience, Theory, and History in Ranke," *History and Theory* 14, no. 4 (1975): 1–14; Peter Burke, "Ranke the Reactionary," in *Leopold von Ranke and the Shaping of the Historical Discipline*, eds. Georg G. Iggers and James M. Powell (Syracuse: Syracuse University Press, 1990), 41. Mario Wimmer, *Archivkörper: Eine Geschichte historischer Einbildungskraft* (Konstanz: Konstanz University Press, 2012), 250–271, provides fascinating context surrounding Ranke's fetishism of the archive.
8 Karl Popper, *The Poverty of Historicism* (New York: Harper, 1961), 3.
9 Beiser traces historicism's emergence in a philosophical tradition that began well before Hegel and more proximally relates to Montesquieu and Herder, see Frederick C. Beiser, "Hegel's Historicism," in *The Cambridge Companion to Hegel*, ed. Frederick C. Beiser (Cambridge: Cambridge University Press, 1993), 270–300.
10 Frederick C. Beiser, *The German Historicist Tradition* (Oxford: Oxford University Press, 2011), 1–26.

on earth over time and in epochal stages, rather than with how, as in Kant, we subjectively *view* the course of history as unitary according to the idea of progress.[11] Hegel's portrayal of world history as unfolding toward its final goal in effect translates Kantian subjective purposiveness into the actual, ultimate connectedness of the world and its history. Hence Marcuse argues that teleology belongs to the ontological character of being itself in Hegel's thought.[12] The *Aufhebung* that accomplishes the dialectical connection between opposed moments, elements, and historical tendencies constitutes a historicizing solution to what Hegel represents as the stasis of Kantian theories of cognition and metaphysics. Hegel's well-known dictum "The true is the whole" reflects his dedication to a holistic, unified account of world history, but also refers to the dynamic unfolding of that whole.

Marjorie Levinson's 2007 articulation of a "New Formalist" approach to literary studies has defined historicism more broadly as encompassing a range of approaches to literary and historical phenomena, in contrast to approaches that focus on formal and aesthetic elements.[13] This broad definition echoes in certain respects Emil Fackenheim's description of *Historismus* as treating all philosophical questions as if they can be superseded by historical ones.[14] Historicist approaches to literature, as Levinson notes, are widely seen as most conducive to ideology critique, materialist criticism, and activist approaches.[15] That is, historicizing approaches are viewed as politically and morally more sensitive and more useful. In twentieth-century literary-theoretical debates, formal, aesthetic, and deconstructive approaches to literature have been characterized as politically irrelevant, dangerous, or otherwise morally questionable, owing to their preoccupation with the autonomy and distinctness of the literary work. This worry about critical impotence can be understood as a long-term consequence of Jakobson's identification in 1921 of "literariness" as the proper object of critical investigation.[16]

---

11 Cf. Rudolf Makkreel, "Purposiveness in History: Its Status after Kant, Hegel, Dilthey, and Habermas," *Philosophy & Social Criticism* 18 (1992): 225.
12 Herbert Marcuse, *Hegel's Ontology and the Theory of Historicity*, trans. Seyla Benhabib (Cambridge: MIT Press, 1987), 139–140.
13 Marjorie Levinson, "What Is New Formalism?" *PMLA* 122, no. 2 (2007): 558–569.
14 Emil Fackenheim, *Metaphysics and Historicity* (Milwaukee: Marquette University Press, 1961). See also Karel Kosík, "Historism and Historicism," *New German Critique* 10 (1977): 65–75.
15 Levinson, "What is New Formalism?," 4. Joseph North explains the historicist turn in literary studies, while it may appear to be in service of progressive aims, as a neoliberal phenomenon; see *Literary Criticism: A Concise Political History* (Cambridge: Harvard University Press, 2017).
16 Roman Jakobson, "Modern Russian Poetry: Velimir Khlebnikov," in *Major Soviet Works: Essays in Criticism*, ed. E.J. Brown (Oxford: Oxford University Press, 1973), 62.

Nevertheless, the nineteenth-century German philosophy of history does not entirely support such a dichotomization between formalism and historicism.[17] To be sure, even Hegel's philosophy of history can be read as confounding this dichotomy. Although Popper lambastes Hegel as "the source of all contemporary historicism" and Hegel condemns Kantian formalism, there is nonetheless a profoundly formal quality in Hegel's dialectic.[18] As John McCumber notes, Hegel's *Phenomenology* presents "the 'formal' side of Spirit's rise."[19] The *Phenomenology* depicts several *forms* of consciousness as remaining the same even when their *objects* differ, such that those forms recur in increasingly complex and contradictory ways. Sense-certainty, for instance, is immediately certain of its own "this," its own object, and contradictions unfold that are recapitulated at higher levels of spirit, for instance when moral conscience is immediately certain of its duty. In each case, the particular form of certainty turns out to have a dubious and mediated relationship to that of which it is certain. The structure, motif, or form of a certainty inhabited by an uncertainty recurs at each level in increasingly complex ways. A narrower example of formalism in Hegel appears in the lectures on the philosophy of history: Hegel notes that a human figure kneeling in prayer takes the same form whether the object of prayer is an idol or the Christian God.[20] The form – here the actual configuration of the human body – remains the same even when the object of worship differs. This passage, although its scope is minuscule compared to the project of world history, concretizes the significance in Hegel of the recurrence of forms. Although Hegel explicitly rejects Kantian moral formalism, his historicism incorporates a profound formal consistency, insofar as forms of human life, and their contradictions, repeat and recur.

Ranke's constructed historicism

With regard to Ranke, the historicist label obscures the degree of attention he accords to formal and aesthetic issues in historiography. Perhaps most important

---

17 See for instance *Representations* 104 (2008), a special issue on form, in particular Catherine Gallagher, "The Formalism of Military History," 23–33; Carla Hesse, "A Fugitive Book," 37–49; Thomas Laqueur, "Form in Ashes," 50–72; Samuel Otter, "An Aesthetics in All Things," 116–125; and Randolph Starn, "Historicizing *Representations:* A Formal Exercise," 137–143.
18 Karl Popper, *The Open Society and its Enemies* (London: Routledge, 2011), 246.
19 John McCumber, *The Company of Words: Hegel, Language, and Systematic Philosophy* (Evanston: Northwestern University Press, 1993), 143.
20 Hegel, *Lectures on the Philosophy of World History, Introduction: Reason in History*, trans. H.B. Nisbet (Cambridge: Cambridge University Press, 1975), 45.

in this context is Hayden White's argument for Ranke's aesthetic contributions to nineteenth-century historiography. White writes that "[Humboldt and] Ranke held that history is ultimately [...] a *mimetic* art form concerned with the representation of reality."[21] White characterizes Ranke as an inventor of a new genre of "historical realism" at the intersection of literary and historiographical production and in tune with the literary realism of the nineteenth century.[22] As White and others have suggested, Ranke's conception of history depends on the rejection of idealist philosophy, mechanism, positivism, religious dogmatism, Romantic art, and positivist science.[23] Ranke's historicism is not antiaesthetic, ultrapositivist, nor triumphalist, despite a legacy that represents it as such, including Benjamin's famous reference.[24]

Ranke's methodological considerations of historiography in fact contain explicitly aesthetic reflections. Ranke extensively interrogates the constructive task of the historian, recognizing that artfulness and artifice are required by both. Ranke states in his lecture "On the Relation and Distinction between History and Politics" that "history is based wholly on literature."[25] He also asserts in his review of Davila's *History of the French Civil Wars* that

> Die Aufgabe des Historikers [...] ist zugleich literarisch und gelehrt. Die Historie ist zugleich Kunst und Wissenschaft. [...] [Sie] soll dem gebildeten Geiste denselben Genuß gewähren wie die gelungenste literarische Hervorbringen. Man könnte sich zu der Annahme neigen, als ob die Schönheit der Form sich nur auf Kosten der Wahrheit erreichen lasse. [...] Ich halte mich jedoch von dem Gegenteil überzeugt und denke, daß das auf die Form gerichtete Bestreben sogar den Eifer der Untersuchung befördert. (*SW* 12:5)

> The task of the historian [...] is at once both learned and literary, for history is at once both art and science. [...] It must [...] offer the educated reader the same pleasure as the most accomplished literary production. One might perhaps incline to the belief that beauty of form can be attained only by a sacrifice of the truth. [...] But I am convinced of the contrary. I think that the effort to improve the form also improves the investigation.[26]

---

21 Hayden White, *Metahistory: The Historical Imagination in Nineteenth-Century Europe* (Baltimore: The Johns Hopkins University Press, 1973), 187.
22 White, *Metahistory*, 166.
23 White, *Metahistory*, 164.
24 See Pieter Geyl, *From Ranke to Toynbee: Five Lectures on Historians and Historiographical Problems* (Northampton, Mass.: Smith College, 1952).
25 Ranke, "On the Relation of and Distinction between History and Politics," in *Theory and Practice of History*, 80.
26 Ranke, "The Historian's Task," in *The Secret of World History: Selected Writings on the Art and Science of History*, ed. and trans. Roger Wines (New York: Fordham University Press, 1981), 258.

Ranke shows himself in these instances to be entirely sensitive to the aesthetic, formal, and literary aspects of historiography. Ranke enunciates, in other words, a dedication to artistic, aesthetic, and formal excellence in historiography, which is obscured in the characterizations by Benjamin, Simmel, and Adorno, and also by the association of Ranke with positivism.

Ranke's treatment of the writings of Renaissance historian Francesco Guicciardini is instructive in view of Benjamin's later dismissal of "how it really was." For in Ranke's commentaries on Guicciardini he reflects upon the weaknesses and strengths of hodgepodge historical narratives. Ranke first disparages the fragmentary aspects of Guicciardini's writing:

> [D]iese Geschichte [ist] [...] zum guten Theil aus anderen Büchern, ohne besondere Forschung, zusammengetragen. Ein großer Theil derselben, Reden, [sind] keineswegs historische Monumente, sondern Uebungen der Redekunst [...] wichtige Facten [sind] ganz entstellt. (*SW* 34:37)

> [T]his History [...] is to a large extent put together from other works, without any particular research; [...] a great part of it, the speeches, is in no way an historical monument, but an exercise in oratory; [...] important facts are completely distorted.[27]

Ranke observes that Guicciardini's history is not cohesive; it is rather an untheorized collage of quotes drawn from other works – an oratorical, hence aesthetic, work rather than a historical one. The resonance with Benjamin's *Passagenwerk* is notable. Nonetheless, Ranke ends up actually praising the form of the work, pronouncing Guicciardini successful *thanks to* the disruptive commentary and digressions:

> Da kamen denn die Discorse Guicciardini's, diese Betrachtungen jeder Begebenheit, von allen Seiten, zur rechten Stunde. [...] Man fühlte sogleich, daß dies die Hauptsache in dem Werke sei. [...] Man muß gestehen, diese Discorse in Guicciardini etwas wahrhaft Originales, daß sie voller Geist und Scharfsinn sind. (*SW* 34:37)

> [T]here are the discourses of Guicciardini, his considerations regarding every event, at the right moment . [...] One feels [...] that this is the most important aspect of Guicciardini's work. [...] [T]hese discourses [...] have something original in them, full of spirit and critical sense. ("Critique of Guicciardini," 97)

Here Ranke lauds Guicciardini's digressive comments for their effect and even their opportune placement within the text. He adds, "For the purposes of presenting these ideas, it would not have been right for Guicciardini to relate the

---

27 Ranke, "Critique of Guicciardini," in *Secret of World History*, 96, translation modified.

major events without interruption by lesser ones. Each was as important as the other."[28] Ranke thus *lauds* the interruptive style and the incongruity of different magnitudes of events in his narratives. The interruption and mixing of elements is deemed successful and for this reason, Guicciardini's formal decisions are praised.

Ranke approaches Guicciardini with special focus on the formal and artful qualities that evoke particular effects. It is therefore *not* as an account of "how it really happened" that Ranke ultimately finds Guicciardini's work praiseworthy. We even see in Ranke's praise of Guicciardini a proximity to Benjamin, whose interruptive materialist historiography draws out a critical response at a present moment in light of another, unconnected moment. Ranke here portrays the task of history-writing as a matter of presenting things so as to provoke *Scharfsinn*, or "critical sense."[29] That *Scharfsinn* emerges in the context of a striking juxtaposition, a spirited critical selection and arrangement on the part of the historian. These formal choices turn out to be key to the historian's enterprise.

Not progress, not development

Benjamin's invocation of Ranke alongside progressivist ideology is, of course, not unfounded. For instance, in his preface to the first edition of *Histories of the Latin and Germanic Peoples* (1824) Ranke states that the second requirement for writing history, after the strict presentation of facts, is "the development of the unity and the progress of the events" in order to seek "the general line of [...] development [of every people, every power, every individual], the direction they took, the ideas which motivated them."[30] Here Ranke appears to endorse a progressivist narrative with a unifying linear trajectory. He writes that history "must not allow itself to be dissipated in the multiplicity of events, but must press on with steady gaze toward its final goal."[31] That progressivism is radically limited, however, when it comes to the continuity and connections between historical epochs that are supposed to be equally immediate to God. Ranke writes, "[t]he concept of progress [...] cannot be applied to the connections between the

---

[28] Ranke, "Critique of Guicciardini," 98.
[29] To be sure, Ranke also argues for the importance of the historian's "eye for the universal," see "History and Philosophy," in *Secret of World History*, 103.
[30] Ranke, "Preface to *Latin and Germanic Peoples*," 86.
[31] Ranke, "On the Relation of and Distinction between History and Politics," 78. Along these same lines White points to the organicist and mechanical metaphors upon which Ranke relies and to his use of what White calls the comic mode of historical emplotment (White, *Metahistory*, 168).

centuries in general."³² This is the kernel of Ranke's historicism, i.e., the radical *dis*connection of epochs in his thought, which opens him to the charge of relativism. While Ranke echoes the ambitions of universal history at particular moments, nonetheless he discounts the continuity of epochs, emphasizing singularity: "[The value of every epoch] consists, not in what follows it, but in its own existence, its own proper self [...] Every epoch must be regarded as something valid in itself."³³ We see in his attention to the uniqueness of each epoch, rather than the place of each epoch in a continuum of historical progress, Ranke's debts to Herder and the Romantic criticism of universal history.³⁴ He rejects the holism and continuity of universal history in favor of differentiating, multiplying, and diversifying the objects of history, highlighting their differences and disconnections rather than eliding them. Benjamin's contemporary Friedrich Meinecke describes Ranke as even obsessed with particularity, a portrayal that grants Ranke a place in a postmodern lineage of concern with what Fredric Jameson in recent work calls "aesthetic singularity."³⁵

Although, as Beiser shows, Ranke shares some of Hegel's methodological principles, nonetheless he rejects a view of history as the development of Spirit in and through human beings.³⁶ Likewise, Ranke argues against Fichte's five-stage progressivist world plan, pointing out that philosophers dispute the dominating ideas of any one epoch and that there exist peoples who exhibit circumstances outside Fichte's formula.³⁷ Ranke insists upon disconnection and distinctness, on "every epoch" in its own right. As Hajo Holborn writes, Ranke emphasized "the uniqueness of the individual or of an individual period of history" in contrast to what Holborn defines as "philosophies of history [that] postulated a predetermined goal of history and maintained that all individual events were only illustrations of one and the same scheme of historical evolution."³⁸ Ranke's dictum "Every epoch is immediate to God" epitomizes the embrace of singularity. The concentration on the individual event or phenomenon typifies

---

32 Ranke, "Epochs of Modern History," in *Secret of World History*, 162.
33 Ranke, "Epochs of Modern History," 159.
34 Those debts do not include hilarity. The witty title of Herder's 1774 treatise is *Auch eine Philosophie der Geschichte zur Bildung der Menschheit: Beitrag zu vielen Beiträgen des Jahrhunderts* (Frankfurt am Main: Suhrkamp, 1967).
35 Fredric Jameson, "The Aesthetics of Singularity," *New Left Review* 92 (March–April 2015): n.p. Jameson argues that postmodernity's aesthetic is one of singularities, which display a temporality of the present and a resistance to norms, laws, and nominalisms.
36 See Frederick C. Beiser, "Hegel and Ranke: A Re-Examination," in *A Companion to Hegel*, eds. Stephen Houlgate and Michael Baur (Oxford: Wiley-Blackwell, 2011), 332–351.
37 Ranke, "On the Relations of History and Philosophy," in *Theory and Practice of History*, 5.
38 Hajo Holborn, *History and the Humanities* (Garden City, New York: Doubleday, 1972), 93, 94.

Ranke's rejection of Kantian universal history, with its emphasis on the perfection of human nature, and also Ranke's rejection of Hegelian teleology. As Kittsteiner has argued, the dictum even places Ranke in a peculiar proximity to Benjamin.[39] Ranke's focus on an event or epoch in its specificity constitutes a proto-Benjaminian gesture in the legacy of Herder, rather than a point of allegiance to a Hegelian model of connected historical stages.

Ranke and ideology

In Benjamin's wake, Ranke appears as a naïf of theory of history, and historicism appears as the naïf of historiography. The suggestion here, in line with Kittsteiner and others, is that in his formal prescriptions for the writing of history Ranke can be approached instead as a Benjaminian ideology critic *avant la lettre*. Benjamin's historical materialism highlights singular moments rather than presenting a linear narrative and emphasizes the construction of constellations rather than the depiction of intrinsic connections between moments and events. It strives explicitly to explode a concept of progress that, as we have seen, grants universal history a unity. Ranke's reflection on the purposes of writing history hints at a peculiarly similar perspective. Let us recall the also-famous lines that precede the "how it really was": "To history has been given the function of judging the past, of instructing men for the profit of future years. The present attempt does not aspire to such a lofty undertaking."[40] Ranke portrays his own attempt at historiography as diverging from the conventional operations of the historian, i.e., as resisting moralizing narratives and prescriptions for the future. In other words, Ranke's effort to present "how it really was" rejects a presentation of the past according to a specific agenda in which the connections made between past and present would serve ulterior purposes. Indeed, for Koselleck Ranke represents a post-1800 shift away from a didactic, moralizing use of history. In his semantic-hermeneutic approach Koselleck points, for instance, to Ranke's increasing use of the past tense, which evokes the uniqueness of the past rather than its exemplarity for the future.[41]

---

39 Kittsteiner, "Benjamin's Historicism," 185.
40 Ranke, "Preface to *Latin and Germanic Peoples*," 86.
41 Reinhart Koselleck, "Historia Magistra Vitae: The Dissolution of the Topos into the Perspective of a Modernized Historical Process," in *Futures Past: On the Semantics of Historical Time*, trans. Keith Tribe (Cambridge: MIT Press, 1985), 31–32. Also relevant here are Jameson's arguments about the tense of realist novels and what that issue offers to progressive thought. That is, Jameson disagrees with Sartre's dichotomy between on the one hand the preterite

Ranke is likewise not unmindful when it comes to traditional historiography's sympathy for the victors of history, as when he writes "Every century has the tendency to consider itself the most progressive, and to measure all others according to its own ideas."[42] He articulates a demand that history be written in its moment, without regard for past representations of an epoch: "[In modern history] [f]acts which later come to light reveal the attempted reconstruction as erroneous" and hence it is important that the historian "put together a work which does not bear the imprint of the past."[43] Reception, in other words, must be filtered in order to present "how it really was." The present account must be *dis*connected from interpretations that have come before, and the historian must not distort the past in order to render the present its superior. While the desire for a vision of "how it really was" may seem naïve, Ranke proposes it precisely as an epistemological strategy for removing the distortions that past historians have interpreted into their material. The goal of presenting "how it really was" in Ranke is exactly meant not to instruct nor to confirm any particular view at all. It is in contemporary terms radically anti-ideological. While Herder's philosophy of history likewise rejects Kantian universalism, Ranke's anti-Romantic turn to what White defines as a new genre of historical realism sharpens that anti-ideological and anti-triumphalist thrust.

We have seen that Ranke shares certain epistemological interests with Benjamin regarding the task of historiography. There is also an aesthetic and formal aspect of Ranke that carries Benjaminian echoes. Ranke even exhibits a proto-Benjaminian appreciation for the bare, untheorized presentation of a historical moment in his reflections on Guicciardini. He writes, "Guicciardini [...] presents a pure consideration of an object which lies before us."[44] Ranke seems to claim that Guicciardini simply presents the historical moment in question in its singularity – "quotes" it in Benjamin's sense. Ranke also presages Benjamin when he argues for the significance of fragments and for relinquishing guiding ideas when he writes of his own book,

---

tense, closed temporality, and a destinal, irrevocable trajectory of the récit, and on the other hand the open, existential present of freedom Sartre believes to be proper to the novel. Jameson attempts to argue that realism lies at the intersection of these two literary temporalities, i.e. destiny and the eternal present. Insofar as these two converge in realism and are always in tension, Jameson claims that there can be no present realism, only realism in its emergence and breaking down. See Fredric Jameson, *The Antinomies of Realism* (London: Verso Books, 2015), 18.

42 Ranke, "Power and Spiritual Force in History," in *Secret of World History*, 257.
43 Ranke, "The Historian's Task," 258.
44 Ranke, "Critique of Guicciardini," 97.

Freilich ist nun dieses Buch Fragment an Fragment [...] aber der Weg der leitenden Ideen in bedingten Forschungen ist eben so gefährlich, als reizend [...] Etwas Vollkommenes darf also hier Niemand erwarten.

Truly this book represents just one fragment of another fragment [...] the role of guiding ideas [leitende Ideen] in such a limited work is as dangerous as it is attractive [...] Thus no one should expect to find here something complete.[45]

Even Benjamin's favored topos of collection appears in Ranke's introduction, when he compares the "manifold monuments of modern history" – i.e., historical writings – to a great collection of antiquities "gathered from many countries and many ages, and placed next to one another in disorder."[46] In this respect Ranke anticipates Burckhardt's treatment of the collection – which can be seen as a key to Benjamin's development of the dialectical image.[47] Ranke's devotion to singularity and "how it really was" is, not unlike Benjamin's materialist historiography project, meant to strip away projections, justifications, and mythologizations.[48] As we have seen, Ranke rejects universal history; his embrace of singularity leads to a valuation of interruption, incongruent events brought together, and other elements that are not at all far away from Benjamin's own materialist historiography. Hence Ranke writes, "we recognize [...] that human affairs are neither guided by a blind, inevitable fate, nor steered by false visions," a declaration that reflects his critical attitude toward teleology and determinism.[49]

---

45 Ranke, *Zur Kritik neuerer Geschichtsschreiber*, vols. 33–34, *Sämtliche Werke* (Leipzig: Duncker und Humblot, 1885), IV–V; Ranke, "Critique of Guicciardini," 76. For a detailed account of aesthetics and formalism in Ranke and others, see Rudolf Vierhaus, "Historiography between Science and Art," in *Leopold von Ranke and the Shaping of the Historical Discipline*.
46 Ranke, "Critique of Guicciardini," 74. As noted above, he writes that his own book has not managed to be a "historical monument" but instead "an exercise in the art of oratory" (96).
47 See Paul R. Sweet, "Wilhelm von Humboldt (1767–1835): His Legacy to the Historian," *The Centennial Review* 15, no. 1 (1971): 27; this observation is also made by Howard Caygill, "Walter Benjamin's Concept of Cultural History," in *The Cambridge Companion to Walter Benjamin*, ed. David S. Ferris (Cambridge: Cambridge University Press, 2004), 74–77.
48 Max Pensky, "Contributions Toward a Theory of Storms: Historical Knowing and Historical Progress in Kant and Benjamin," *The Philosophical Forum* 41, nos. 1–2 (2010): 170, points out that progress per se is not Benjamin's target, but instead the *idea* of progress, whose destruction would constitute a step of actual historical progress.
49 Ranke, "On the Relation of and Distinction between History and Politics," 82. Another proto-Benjaminian aspect of Ranke's writing is suggested by Krieger, who shows that for Ranke in times of crisis, "politics, thus mediated through contemporary history, became compatible with past history in a way that transcended the normal problem of their relations" (Krieger, "Elements of Early Historicism," 2). Ranke proposes, in other words, that unexpected connections may appear between the past and the present in a moment of crisis – one of

Although Benjamin takes Rankean historicism as the precise opposite of materialist historiography, there are, as we have seen, surprising points of proximity between Ranke and Benjamin. It is noteworthy, moreover, that some of Ranke's worries with regard to his own formal prescriptions for depicting "how it really was" can be regarded as strangely close to what Adorno criticizes in Benjamin's essay on Baudelaire, which was to serve as an exposé for the *Passagenwerk* as a whole. Adorno complains in correspondence about Benjamin's "asceticism," namely the lack of embeddedness and theoretical explanation of his material; Adorno criticizes Benjamin's "wide-eyed presentation of mere facts."[50] Adorno then criticizes the *Passagenwerk* as containing "motifs [that] are assembled but they are not elaborated."[51] For Adorno, Benjamin's spare presentation of quotes and facts reflects insufficient dialectics.[52] Benjamin needs, in Adorno's view, to embed the motifs and facts in a theory, so as in effect to narrativize them. A century before Benjamin writes "On the Concept of History," when Ranke asks himself, "What will be said of my harsh, disconnected presentation of history?," he conjures a criticism that turns out to resemble the Adornian complaints against Benjamin. Ranke promulgates a proto-Benjaminian defense against this anticipated objection: "Strict presentation of facts, no matter how conditional and unattractive they might be, is undoubtedly the supreme law."[53] Ranke goes on to add, "A lofty ideal does exist: To grasp the event itself in its human comprehensibility, its unity, and its fullness."[54] The spotlighting of the event in its individuality resembles once again the construction of constellations in Benjaminian materialist

---

Benjamin's well known formulations – connections which diverge from the relationships historians generally examine. This sounds like a rejection of the causal relationship, and even like a reference to the citability that forms the basis of Benjamin's constellation of historical moments. Beatrice Hanssen, *Walter Benjamin and the Arcades Project* (London: Bloomsbury, 2006), 24, argues that Benjamin implicitly refers to Ranke when he evokes the citability of the past in all its moments in the third thesis of "The Concept of History."

50 Walter Benjamin and Theodor Adorno, *The Complete Correspondence, 1928– 1940* (Cambridge: Polity, 1999), 283.
51 Benjamin and Adorno, *The Complete Correspondence*, 281.
52 See my "Not Dialectical Enough: On Benjamin, Adorno and Autonomous Critique," *Philosophy and Rhetoric* 44, no. 4 (2011): 336–362.
53 Ranke, "Preface to *Latin and Germanic Peoples*," 86. Let us also note that Ranke's criticism of the disconnected, collage-like quality of Guicciardini's account bears a similarly striking resemblance to Adorno's criticism of Benjamin's draft outline for the *Passagenwerk*.
54 Ranke, "Preface to *Latin and Germanic Peoples*," 87.

historiography, which should dislodge an event from a narrative that is based on a stream of causality and bring that event into its disruptive and constellative potential vis-à-vis other moments and resonances.

Old and older historicism

There is an intersection of formal and epistemological questions in the proleptically Benjaminian character of Ranke's approach to historiography. These include Ranke's sensitivity to the formal arrangement and connection of elements, his emphasis on the singularity of moments, his rejection of didacticism, his appreciation of interruptive elements and, as Burke has designated it, his "react[ion] against an earlier historical revolution."[55] In certain respects both Benjamin and Ranke reject their predecessors as naïve and as committed to different versions of triumphalism. Ranke explicitly places himself in a lineage in which he is an innovator, a developer of a new historicism that is preoccupied with particulars: "The error of the philosophers of the last century was that they formulated a universal doctrine [...] They avoided the painstaking work [...] by which particulars could be discovered."[56] Ranke sees himself, in other words, as an innovator, reacting against a tide of universal history.

What is more, Ranke's embrace of the singular represents, even for Ranke himself, an explicit rejection of the interpretations of his own historiographical predecessors. An 1873 letter to his son Otto Ranke explains that rejection: "[T]he running current [in historical scholarship] seeks to dominate the past and force it into its course."[57] Ranke represents himself as breaking with a form of historiography that "dominates" and "forces" the past into a narrative, freeing the particulars from a coercive and even triumphalist context. He portrays himself as swimming against "the running current" in historiography, as rejecting what we might call an ideological effort to mold a historical account. Ranke's historiographical project renders him a self-consciously newer sort of historian, in comparison to what he calls the historical "monuments," i.e., the classic histories:

> [Einige] wollen darstellen: es dünkt sie der Weg der Alten, den sie nehmen. Andere wollen aus dem Vergangenen Lehren für die Zukunft herleiten; Viele wollen vertheidigen oder anklagen; nicht Wenige bemühen sich, die Begebenheiten aus tieferen Gründen, aus

---

55 Burke, "Ranke the Reactionary," 37.
56 Ranke, "On the Relation of and Distinction between History and Politics," 116.
57 Ranke, "The Historian's Ideal," in *Secret of World History*, 259.

> Gemüth und Leidenschaft zu entwickeln. Dann sind Einige, die nur den Zweck haben, zu überliefern, was geschehen ist. (*SW* 34: iii)
>
> [Some] pretend to represent the ways of the ancients. Others seek to draw from the past lessons for the future. Many wish to defend or accuse. Not a few strive to develop the events out of deeper motives of emotion or passion. Then there are a few whose only purpose is to transmit what happened. ("Critique of Guicciardini," 74–75)

Ranke here provides evidence for Koselleck's claim that historiography after 1800 turned from a task of teaching eternal lessons to a linear presentation of singular events. Ranke's historiography proposes not to find lessons, nor deeper motives, but simply to convey "how it really was." The transmission of "how it really was" is counterposed to the moralizing tendencies of history hitherto. It constitutes in that respect a rejection of what critical theory might define as ideological interests. In the wake of Benjamin, the phrase is taken as a marker of naïveté and associated with triumphalism, positivism, and the victory of the winners. For Ranke, however, it signals a rejection of the universal historians' characterization of laws that govern the connections between events and epochs – a rejection of triumphalism and progressivism that Benjamin in his historical materialism performs otherwise. The teleological, linear-progress model that Ranke imputes to Hegel, and together with which Benjamin implicitly lumps Ranke, eclipses the uniqueness of each epoch in connecting them by laws and principles of progress and teleology. Ranke's historicism includes an aesthetic, creative, particularizing, and subjective approach – and yet Benjamin takes Ranke as an exemplar of the historiography to which Benjamin opposes his own particularizing, subjective, and aesthetically inflected approach.

Ranke thus rejects universal history and the Hegelian portrayals of a law-governed unfolding of *Geist*. He also rejects historiographical predecessors who might be seen as proto-positivist. In an introduction to a lecture on universal history, Ranke explicitly opposes himself to the "old traditional method" of history which seeks primarily "to establish time, place, sequence."[58] The "old" theories of history, Ranke writes, seek to include only the factual elements, but Ranke notes that with them, "the mass of facts are not easily followed; the overall impression is infinitely bleak."[59] He adds that their approach leaves out "for what purpose all these things happen [...] even the inner connection is

---

[58] Ranke, "The Pitfalls of a Philosophy of History," in *Theory and Practice of History*, 17.
[59] Ranke, "Pitfalls," 17.

distorted."[60] For Ranke, a proto-positivist, "just the facts" approach lacks a principle of connection and coherence. Likewise Ranke differentiates himself from previous historians when he writes that he is interested in the absolute coincidence between *Geschichte*, the objective relationship to "what happened," and *Historie*, which he defines as "the more subjective relationship."[61] Ranke intimates that both Hegel, with his view of history as the development of Spirit in and through human beings, and the "old" proto-positivist methods of historiography each overlook the significance of individuals and of subjectivity. The Hegelian history forces "how it really was" into a law-governed narrative, and the proto-positivist concentration upon facts alone does not provide enough narrative, i.e., enough connection.

Ranke's much-quoted demand that historians portray "how it really was" is less a naïve statement of unmediated certainty than it is an explicit contrast to accounts that are driven by laws that serve moralizing purposes or ideology. Ranke sets himself up as a cautious, partial, fragmentary author of a history composed of relatively bare material – i.e., disembedded and untheorized – and yet with a sensitivity to the "subjective relationship" to events. This is not to exaggerate the Benjaminian elements of Ranke nor to intimate that Ranke offers a materialist historiography, or uses dialectical images, about which Benjamin writes in such cryptic fashion. On the other hand, Ranke does not seem truly to believe that historiography involves the representation of the past "how it really was," without intervention on the part of imagination, subjectivity, and something like aesthetic experience. Leonard Krieger has described a "mutual autonomy between Ranke's transcendental faith and his passion for particular historical truths."[62] In Meinecke's 1936 "Die Entstehung des Historismus," Meinecke characterizes Ranke as absolutely devoted to individualities, which, as Krieger points out, understates Ranke's devotion to principles of historical universality.[63] Nonetheless, while Krieger reminds us of Ranke's commitment to generalities, he acknowledges that it is not easy for Ranke to maintain a consistent fidelity to them.[64] Ranke seems on the one hand to cherish the singular while also demanding narrative coherence.

---

60 Ranke, "Pitfalls," 18.
61 Ranke, "Pitfalls," 19.
62 Krieger, "Elements of Early Historicism," 1.
63 Krieger, "Elements of Early Historicism," 4–5.
64 Krieger, "Elements of Early Historicism," 6.

This raises the question of why Ranke, with his insistence on subjective experience and differences, is seemingly lumped together by Benjamin with linear, triumphalist narrative, while Ranke contrasts himself to older historiography and rejects, in a Nietzschean vein, earlier uses of history. Is there a trope enacted here, in which serial generations of historians decry their predecessors as traditional historians, in order to propose something more imaginative, more creative, more disruptive, and less in thrall to teleology?[65] It is worth noting that "wie es eigentlich gewesen" is not new with Ranke. Peter Burke has shown that this phrase goes back at least as far as 16th century historians Johannes Sleidanus ("prout quaeque res acta fuit") and Lancelot Voisin de La Popelinière ("reciter la chose comme elle est avenue").[66]

We have seen that Ranke embraces disruption, discontinuity, and absolute singularity. He relinquishes guiding ideas, praises interruptions that promote *Scharfsinn*, and endorses the creative production of narrative connection by way of inference and imagination. Hence even for such an "old historicist" as Ranke, writing history requires literary capacities; in effect it demands submission to an aesthetics of historical form along with an oscillation among the various possibilities for its requirements. Ranke is painted by Benjamin as the old historicist against whom materialist historiography is a reaction. But Ranke opposes himself to older methods – both proto-positivist and Hegelian.[67] Ranke objects to law-governed historiography in favor of the singularity of an event or epoch and also insists on the embedding of singularities in a narrative. He prefigures the tension between Benjamin and Adorno regarding Benjamin's disconnected treatment of historical singularities and the demand for narrative and theorization. The reception of Ranke, however, in particular in Benjamin's

---

[65] Here it is worth considering the nuanced view of Nietzsche's responses to nineteenth-century German historicism in Thomas Brobjer, "Nietzsche's Relation to Historical Methods and Nineteenth-Century German Historiography," *History and Theory* 46, no. 2 (2007): 155–179. Although Nietzsche's condemnations of Ranke and historicism are well known, Brobjer shows that there are substantial hints of ambiguity in that relationship. For instance, while Nietzsche dismisses Ranke in his famous second untimely meditation, he actually writes that Ranke opened his eyes to history (Brobjer, "Nietzsche's Relation," 171–172). Brobjer argues that the essay is not typical of Nietzsche's assessments of nineteenth-century historical writing and that Nietzsche repudiated those earlier assessments (Brobjer, "Nietzsche's Relation," 156). Brobjer also notes that Nietzsche had at least some temporary sympathy with followers of Ranke including Heinrich von Sybel, together with J. G. Droysen and Heinrich von Treitschke (Brobjer, "Nietzsche's Relation" 175).

[66] Burke, "Ranke the Reactionary," 192n2.

[67] Ranke, "Pitfalls," 18.

wake, downplays Ranke's own formal and theoretical engagement. His stress on the specificity of different objects and epochs, the effectivity of language and form, and the role of imagination and collection in achieving proper connection or in covering over disconnection bespeak neither a strictly empiricist historical style, nor a triumphalist linearity. Instead, they demonstrate a style of historiography that attends to singularity, disruption, and the formal demands and limitations of representing historical connection.

Part III: **Epochality: On Phenomenology's Appeals to a Disconnected Past**

6 Heidegger and the Plot of Metaphysics

Ranke's emphasis on "every epoch" finds a surprising echo in twentieth-century phenomenology. Insofar as post-Husserlian phenomenology takes on an "epochal" cast in Heidegger and other philosophers indebted to him, the distinctions between epochs and the particular qualities of modernity become topics for phenomenology. There is, as we shall see, nonetheless a tension between the structural and "synchronic," quasi-scientific inquiry into the character of a distinct phenomenon on the one hand and the claims for modernity and the sequence of epochs on the other hand. This is the case with Heidegger, Arendt, and Koselleck, but perhaps most obviously with Heidegger's own shifting definitions of "historicity" itself.

In early essays and in *Being and Time*, Heidegger conducts a phenomenological inquiry into what he calls *Geschichtlichkeit*, or historicity. In these contexts, historicity pertains to Dasein's being-in-the-world, its "how." Historicity takes on a different sense, however, in Heidegger's many formulations of the history of metaphysics. There historicity is associated with the trajectory of western civilization, which is to say, with history per se. Heidegger develops a narrative in which the pre-Socratic experience of truth was obscured in different ways over several epochs, culminating in the present age of technology. The historicity of Heidegger's narrative of western civilization is far removed from the historicity of Dasein's "how." The former involves a series of epochs, connected over time in their modes of forgetfulness, while the latter refers to Dasein's way of being. Although Heidegger's interest in historicity as a quality peculiar to Dasein appears to fade after *Being and Time*, this chapter will argue that there is nonetheless an obscure but important continuity between the *form* of the investigation of Dasein's historicity and the *narrative* of the history of metaphysics. That is, the trajectory of his early phenomenological inquiry, which lays bare Dasein's quasi-essential, temporal movement of disclosure, reappears in narrativized – and hence emplotted – form in Heidegger's history of metaphysics.

We will begin with the observation that Heidegger's phenomenological investigations into historicity – and into other concepts and phenomena including art and truth – display a pronounced arc, procedure, or what we might even call a plot: each begins with a commonsense view or definition; Heidegger then questions that commonplace in a variety of ways, excavating and dismantling a succession of misconceptions; and he ultimately uncovers an originary temporal event that forms the "essence," "truth," or "origin" of the concept or phenomenon at issue. Jean-François Lyotard defines this approach in terms of anamnesis, also claiming that Heidegger attempts to "counterseduce" his audience away from the distortions of everyday language and from the casual usage of words.¹ Heidegger's phenomenology of Dasein's historicity follows just such an anamnestic procedure or plot, uncovering layers of everyday thinking about "self" and "I" in order to expose our eventlike, temporal way of being-in-the-world. A similar anamnestic arc or emplotment of the phenomenological inquiry also defines, for instance, Heidegger's "The Origin of the Work of Art" and "On the Essence of Truth."²

In the following pages, we will see that Heidegger's epochal account of the history of metaphysics *historicizes* that phenomenological procedure. It translates into a unified historical narrative of connected epochs the stepwise, anamnestic trajectory of inquiry that characterizes Heidegger's phenomenologies of Dasein, art, truth, etc. Heidegger in fact adapts the anamnestic procedure of phenomenological inquiry into Dasein's historicity in such a way as to compose an anamnestic, epochal emplotment of western history. History and narrative form are therefore profoundly intertwined here, insofar as the history of metaphysics "epochalizes" the phenomenological inquiry. Heidegger's epochal approach to history, which uncovers an originary experience of truth as *aletheia* that had been forgotten or concealed in the history of metaphysics, converts the emplotted *structure* of his phenomenological investigations into an integral, unified *narrative* about history. To put it another way, the history of metaphysics can be read as allegorizing, along the axis of history, the arc or plot of Heidegger's earlier investigations into Dasein's historicity.³

---

1 Jean-François Lyotard, *Heidegger and 'the jews,'* trans. Andreas Michel and Mark Roberts (Minneapolis: University of Minnesota Press, 1990), 59.
2 Martin Heidegger, "On the Essence of Truth (1930)," trans. John H. Sallis, in *Pathmarks*, ed. Will McNeill (Cambridge: Cambridge University Press, 1998), 136–154; Heidegger, "The Origin of the Work of Art," in *Off the Beaten Track*, trans. and eds. Julian Young and Kenneth Haynes (Cambridge: Cambridge University Press, 2002), 1–56.
3 See for instance Charles Guignon, "History and Historicity," in *A Companion to Phenomenology and Existentialism*, eds. Hubert L. Dreyfus and Mark A. Wrathall (Oxford: Blackwell, 2008), 545–558; Christopher Fynsk, *Heidegger: Thought and Historicity* (Ithaca: Cornell University Press, 1993); Otto Pöggeler, "'Historicity' in Heidegger's Late Work," *Southwestern Journal of Philosophy* 4 (1973): 53–73; Thomas Sheehan, "Geschichtlichkeit/Ereignis/Kehre," *Existentia: An International Journal of Philosophy* 11, nos. 3–4 (2001): 241–251; Thomas E. Wren, "Heidegger's Philosophy of History," *Journal of the British Society for Phenomenology* 3 (1972): 111–125; Jeffrey Andrew Barash, *Martin Heidegger and the Problem of Historical Meaning* (New York: Fordham University Press, 2003); Philip Tonner, "Epoch: Heidegger and the Happening of History," *Minerva: An Open Access Journal of Philosophy* 19 (2015): 132–150; William D. Blattner, "Heidegger and Philosophical Modernism," *Inquiry: An Interdisciplinary Journal of Philosophy* 38, no. 3 (1995): 257–276; Sandra

Dasein's historicity

Heidegger's account of Dasein's historicity appears in condensed form in two early essays, in which Heidegger distinguishes between what we usually understand as history and the historicity of Dasein. In his 1919–1921 response to Jaspers's *Psychology of Worldviews* and in a 1925 essay on Dilthey, Heidegger attempts to reconceive history, i.e., not as an object of study and not as pertaining to a past that would be seen as some particular thing we can know. He pointedly repurposes the word *geschichtlich* in order instead to evoke the sense of the "I" as a way of being.[4] (Although it will be abandoned in the Dilthey essay and *Being and Time* in favor of the word "Dasein," Heidegger utilizes the vocabulary of the "I" in the early Jaspers essay.) The "I" is de-substantialized and refigured in Heidegger's analysis as a temporalized, eventlike "how" rather than a "what." In connecting the "I" to history, reconceived as "happening" rather than as bygone events, Heidegger undermines the sense of the "I"'s identity and constancy. He emphasizes instead its dynamic, eventlike character, what we might call its processual or performative aspect.[5] He portrays the "I" as an "enacting" of its "am," as a "way of being" ("Jaspers," 25). Existence is likewise characterized as "the particular 'how' of the self" ("Jaspers," 25). These portrayals stand in contrast to an objectifying treatment of the "I", what Heidegger calls a "regional" approach, which "turn[s] the 'I' into an object" ("Jaspers," 26).

Heidegger uses similar formulations when he rejects the Husserlian subjectivism of self in *Being and Time*. He explains Dasein's individualization within a phenomenology of "mineness" [*Jemeinigkeit*], while eschewing an orientation toward a self or ego.[6] As in the earlier essays, history, the "I", and existence are desubstantialized, de-essentialized, and rendered in terms of "how" and

---

Lee Bartky, "Heidegger and the Modes of World-Disclosure," *Philosophy and Phenomenological Research* 40, no. 2 (1979): 212–236.

4 Heidegger, "Anmerkungen zu Karl Jaspers Psychologie der Weltanschauungen," in *Wegmarken* (Frankfurt am Main: Klostermann, 1976), 1–44; "Comments on Karl Jaspers' Psychology of Worldviews," trans. John van Buren, in *Pathmarks*, 1–38, hereafter referred to in text as "Jaspers"; Heidegger, "Wilhelm Dilthey's Research and the Current Struggle for a Historical Worldview," in *Becoming Heidegger: On the Trail of His Early Occasional Writings, 1910–1927*, eds. Theodore Kisiel and Thomas Sheehan (Evanston: Northwestern University Press, 2007), 238–274, hereafter referred to in text as "Dilthey."

5 Heidegger's formulation of *Geschichtlichkeit* previews the "performative turn" of poststructuralism, see my "On the Performative Difficulty of Being and Time," *Philosophy Today* 44, no. 4 (2000): 366–379.

6 Heidegger, *Being and Time*, trans. Joan Stambaugh (Albany: SUNY Press, 1996), 41–43. Hereafter cited in text as *BT*.

enactment. Heidegger writes, "The 'historical' is here not the correlate of theoretical and objective historical observation, rather it is both the content *and* the "how" of the anxious concern of the self about the self" ("Jaspers," 28). He recasts history in terms of the dynamic, eventlike "I" rather than as a matter of the past. The Jaspers essay explains that the problem with a theoretical and objective approach to history is that it treats the past as an "appendage that the 'I' drags along with itself" ("Jaspers," 27). Scholarly historiography takes up precisely such a subject-object approach that obscures the true nature of the "I" and of history. What is obscured is a more fundamental or primordial phenomenon of the "I" as a "how," as enactment, and as experience. The essay emphasizes that the "I" *experiences* the past, rather than merely *relating to* bygone events.

The Jaspers essay further explains that the past is not even truly past, for the "I" extends into the past and into the future: "[The past] is experienced as the past of an 'I' that experiences it historically within a horizon of expectations in advance of itself" ("Jaspers," 27). Hence the past comes along with us and even *as* us rather than as an object of study. Similarly, in *Being and Time*, Dasein is described in the mode of "having-been." As having-been, Dasein is represented as fully connected to its past and future, and also to the world, rather than as static and self-enclosed. The present perfect continuous tense of "having-been" reflects this extension as it draws the past into the present. The verb tense evokes the intrinsic, ongoing connection between the past and the present, as opposed to the perfect tense, which is discontinuous and closes off an action or event from the present. This processual, open-ended temporality of the past is one key to Heidegger's recasting of history.

In the course of the Jaspers essay, not only the "I" but also history is redescribed as a *way* we are that includes our past, rather than as an object or situation, now absent or bygone, to which we relate:

> [Die faktische Lebenserfahrung selbst ist] zwar primär nicht ein objektgeschichtliches (mein Leben gesehen als sich abspielend in der Gegenwart), sondern ein sich selbst so erfahrendes *vollzugsgeschichtliches* Phänomen. ("Anmerkungen zu Karl Jaspers, *Psychologie der Weltanschauungen*," 32, *sic* emphasis)

> [This factical experience of life is] not primarily an objective historical phenomenon (my life seen as something that takes place in the present) but rather a phenomenon of *historical enactment* that experiences itself in such enactment. ("Jaspers," 28, *sic* emphasis, translation modified)

Here the "phenomena" of life and history are described once again in terms of temporal happening, enactment, and experience. Likewise Heidegger declares that "the context of [one's] experience also has a historical nature in accord with its sense of enactment" ("Jaspers," 28).

The Dilthey essay covers themes that will be central for Division I of *Being and Time:* being-in-the-world, Dasein's practical concern, being-with-others, inauthenticity, mineness, being futural, and guilt ("Dilthey," 259–262).[7] For the purposes of this chapter what is significant is the emphasis, even greater than that of the Jaspers essay, that is placed on the concept of history. History is characterized as the dynamic, temporalizing enactment of Dasein: "History signifies a happening that we ourselves are" ("Dilthey," 271). Heidegger specifically embraces Dilthey's assertion "that the basic character of life is *to be* historical" ("Dilthey," 270). Nonetheless, when it comes to the concept of life Dilthey obscures the significance of temporality and movement, according to Heidegger's account.[8] Heidegger's own temporalized sense of historicity incorporates Dasein's eventlike quality as well as its being-in-the-world ("Dilthey," 270). In the Jaspers essay, the Dilthey essay, and ultimately in *Being and Time* Heidegger's deployment of *Geschichtlichkeit* will insist on this constellation of qualities.

The emplotment of inquiry

In the Jaspers essay, Heidegger rejects a conceptualization of history in which history is a matter of recorded past events and instead characterizes the historicity of the "I" temporally, as enactment, event, and experience. In the Dilthey essay, the conventional concepts of history, time, and life are recast in a similar way with respect to human Dasein, i.e., as temporalized, enacting, and as a "how." This of course prefigures the central gesture of *Being and Time*, namely insofar as the "what" of Dasein is rethought and reformulated in terms of a temporalized, eventlike "how." A similar form of inquiry recurs throughout

---

[7] See Hans-Georg Gadamer, "Martin Heidegger's One Path," in *Reading Heidegger from the Start: Essays in His Earliest Thought*, eds. Theodore Kisiel and John van Buren (Albany: SUNY Press, 1994); Charles Bambach, *Heidegger, Dilthey and the Crisis of Historicism* (Ithaca: Cornell University Press, 1995); Guillaume Fagniez, "Hermeneutik im Übergang von Dilthey zu Heidegger," *Studia Phaenomenologica: Romanian Journal of Phenomenology* 13 (2013): 429–447; Rudolf A. Makkreel, "The Overcoming of Linear Time in Kant, Dilthey, and Heidegger," in *Dilthey and Phenomenology*, ed. Rudolf A. Makkreel (Lanham: University Press of America, 1987), 41–158.

[8] See Robert C. Scharff, "Heidegger's 'Appropriation' of Dilthey before *Being and Time*," *Journal of the History of Philosophy* 35, no. 1 (1997): 105–128; Eric S. Nelson, "Heidegger and Dilthey: Language, History, and Hermeneutics," in *Horizons of Authenticity in Phenomenology, Existentialism, and Moral Psychology*, eds. Hans Pedersen and Megan Altman (Dordrecht: Springer, 2015), 109–128; Theodore R. Schatzki, "Living Out of the Past: Dilthey and Heidegger on Life and History," *Inquiry* 46, no. 3 (2003): 301–323.

Heidegger's corpus, in particular where Heidegger contorts conventional language in order to evoke the dimension of temporal enactment and event. This linguistic evocation of enactment is striking in such formulations as "World never *is*, but *worlds* [*weltet*]" and "the nothing [...] nothings [*das Nichts* {...} *nichtet*]."[9] These phrasings come across as heteroclite, but such grammatical contortions have a technical name in the rhetorical tradition. The Greek *polyptoton*, related to the Latin *adnominatio*, names a rhetorical device in which a word appears more than once in the same sentence, but as different parts of speech – e.g., "Absolute power corrupts absolutely." Heidegger's deployment of polyptoton in "the world worlds" and "the nothing nothings" also incorporates the trope of *anthimeria*, in which a word is borrowed in order to fulfill the function of a different part of speech – e.g., "I'm all baseballed out, let's go home at the end of the inning."[10]

Despite's Rudolf Carnap's claims that Heidegger's "nonsens[ical]" language invents problems that do not exist, the recognition of such linguistic strategies in the canon of rhetoric signals that Heidegger's formulations are less inventive than they may appear.[11] Heidegger's combinations of polyptoton and anthimeria in such formulations as "the world worlds" highlight the philosophical strains put on grammar, which relies on substantives even when it comes to conveying nonsubstantial events and temporal flux. Heidegger's formulations defamiliarize our understanding of such concepts as "nothing" and "world" in the process of evoking their eventlike character. The polyptoton in that respect accomplishes a defamiliarization, one that Heidegger performs in other ways when he turns from a presentist, static "what" to a temporalized "how."

This turn to the temporalized condition, truth, or essence of an extant phenomenon or seemingly static concept is a hallmark of Heideggerian phenomenology. Many of his lectures and essays follow comparable trajectories. They begin with a significant philosophical concept or phenomenon, such as art or truth; lay out its understanding in common parlance; turn in several steps or stages to ask about its quasi-essence, ground, or origin; evoke as that essence, ground, or origin an eventlike, temporal process that Heidegger characterizes as forgotten, hidden, or concealed; and suggest that the hidden quasi-essence or origin involves an interplay or dynamic relationship between disclosure and

---

9 Heidegger, "What is Metaphysics?" in *Pathmarks*, 90, translation modified; Heidegger, "On the Essence of Ground," in *Pathmarks*, 126.
10 See "Silva Rhetoricae/The Forest of Rhetoric" (website), rhetoric.byu.edu.
11 Rudolf Carnap, "Überwindung der Metaphysik durch logische Analyse der Sprache," *Erkenntnis* 2 (1931): 220–241.

concealment. That interplay cannot serve as a proper essence or ground because as event and enactment it is not a static thing or concept. In addition, it cannot be represented as fixed in the mode of a foundation. For this reason Heidegger also utilizes the language of *Abgrund* – a word play on the German word for "ground" that also denotes an abyss. The *Abgrund* designates a profound lacuna that occupies the position of a foundation.[12] The dynamic event of concealment and unconcealment serves as the essence, condition, or non-foundational *Abgrund*, which is eclipsed by our focus on what is present, static, or conceptually stable.

Heidegger's 1930 essay "The Essence of Truth," to take an important example, begins with the view of everyday understanding or commonsense that truth consists in correctness. Heidegger proceeds to unsettle in several steps the presumptions entailed in that definition of truth and ultimately evokes an event of disclosure and concealment that is most fundamental or essential. "The Origin of the Work of Art" unfolds similarly. It begins with the notion that the origin of the artwork is the artist, dismantles a sequence of misconceptions about art and thinghood, and ends with an event of struggle between earth (on the side of concealment or withdrawal) and world (on the side of disclosure). In each case, the inquiry excavates in stepwise fashion commonplace views that treat the phenomenon (art) or concept (truth) in question as static and object-like. Ultimately Heidegger evokes, as what these objectifying comportments or understandings obscure or forget, the eventlike condition, quasi-essence, or origin of the phenomenon or concept in question. This turn from static phenomenon or concept to a temporalized "essence" is central to the arc of the inquiry in each case. The final move is to expose the eventlike dynamic of concealment and disclosure, whether it is called historicity, the essence of truth, or the strife of earth and world.

The arc or "plot" of phenomenological investigation laid out above – in which the inquiry, in several stages, comes to evoke an eventlike enactment that has been obscured or forgotten and yet forms the essence or quasi-ground of the concept or phenomenon under investigation – also plays out in the anamnestic structure of Heidegger's early inquiries into historicity. The Jaspers essay begins with the concept of "psychical life" and ultimately brings to light

---

[12] See John H. Sallis, "Grounders of the Abyss," in *Companion to Heidegger's* Contributions to Philosophy, edited by Charles E. Scott et al., 181–197 (Bloomington: Indiana University Press, 2001); Parvis Emad, "Translating Heidegger's *Beiträge zur Philosophie* as an Hermeneutic Responsibility," *Studia Phaenomenologica: Romanian Journal of Phenomenology* 6 (2006): 347–368; Daniel O. Dahlstrom, "Abyss/Abgrund," in *The Heidegger Dictionary* (London: Bloomsbury, 2013), 10–11.

the character of the "I am" as a "how" ("Jaspers," 1). The Dilthey essay begins with the concepts of life and "historical worldview" and ultimately exposes Dasein's historicity as a way of being. The "I" or Dasein's being-in-the-world is revealed as an eventlike, in-the-world historicity that forms the unthematized condition of academic historiography. Indeed, the investigations of the historicity of the "I," and of Dasein, the essence of truth, the strife of earth and world, the tool analysis, *Gestell*, and technology all employ this procedure of setting aside or looking beyond everyday, commonsense, or traditional beliefs and opinions. They each then turn their attention to the unthematized disclosure/concealment event that forms the condition or origin of that concept, phenomenon, belief, or comportment. If we wanted to apply to Heidegger his own claim that every thinker thinks only one thought, it would be this anamnestic procedure that constitutes Heidegger's one thought – not being, *Mitsein*, or technology. In each case, as we have seen, Heidegger turns from the phenomenon originally under investigation (art, truth, history) to a forgotten or obscured, more originary, eventlike enactment that constitutes its essence, truth, condition, or possibility, and which consists in a dynamic between disclosure and concealment.

The epochal history of metaphysics

Heidegger's phenomenological approaches turn from commonsense understandings of concepts or phenomena to an anamnestic evocation of the dynamic of disclosure and concealment that forms their eventlike essence, condition, "ground," or *Abgrund*.[13] The main argument of this chapter is that Heidegger's history of metaphysics recapitulates or allegorizes precisely that same anamnestic *procedure*, albeit in the form of an anamnestic *emplotment* that connects historical epochs and unifies the narrative of the forgetting of being. Heidegger's account of the history of metaphysics discovers or remembers a forgotten epoch that is said to have been eclipsed in later epochs and

---

[13] See David Walsh's introduction to Eric Voegelin, *Anamnesis: On the Theory of History and Politics* (Columbia: University of Missouri Press, 2002), 1–27, for a treatment of anamnesis in Heidegger and Voegelin; see also Andrew J. Mitchell, "The Coming of History: Heidegger and Nietzsche against the Present," *Continental Philosophy Review* 46, no. 3 (2014): 395–411; Ned Lukacher, *Primal Scenes: Literature, Philosophy, Psychoanalysis* (Ithaca: Cornell UP, 1986), 45, 109.

its forgetting is what unifies the story.¹⁴ He unfolds a historical narrative in which philosophy forgot the primordial thought and experience of the pre-Socratics, who in contrast *did* experience the event of disclosure as the essence of truth, hence Gabriel Ricci goes so far as to name this a "historiographic anamnesis."¹⁵ In Platonism, according to the narrative, truth came to be conceived as "whatness" with the Platonic idea; under Roman and then Christian thought it was conceived as *veritas* or correctness.¹⁶ The forgotten pre-Socratic epoch, however, experienced the properly eventlike character of the essence of truth. The eclipse of the eventlike character of truth was double – an eclipse of the pre-Socratic epoch as such and of that form of knowledge it experienced. That forgotten, eventlike knowledge-experience underlies the trajectory of the history of metaphysics of which we are a part. It is lost in such a way that truth has come to be understood according to the sole criterion of correctness.

Heidegger's 1937–1938 lecture course, published as *Basic Questions of Philosophy: Selected "Problems" of "Logic"*, includes one variation among many of Heidegger's epochal narrative of the history of metaphysics.¹⁷ That lecture course begins with a cryptic discussion of restraint and "futural philosophy" and then turns to the question of truth in a way that echoes the 1930 essay "On the Essence of Truth." The lectures undermine the commonsense notion of truth as correctness and turn to the essence of truth, understood as freedom or "disclosive letting beings be."¹⁸ Heidegger's excavation of a pre-Socratic experience in the 1937–1938 lectures emplots in epochal terms the same procedure that Heidegger follows over and over in his essays and lectures: he turns to a forgotten "essence" and does so by following a clue or turn of phrase in which

---

14 The reference to epochs in Heidegger is not per se new with the history of metaphysics. At the end of the early essay on Dilthey, for instance, Heidegger refers to "unhistorical ages [*unhistorische Zeitalter*]" as defined by absorption in the present (Heidegger, "Dilthey," 273).

15 Gabriel Ricci, *Time Consciousness: The Philosophical Uses of History* (New Brunswick: Transaction, 2010), 69–70. What Ricci calls "historiographic anamnesis" is not to be confused with the anamnesis described by Lyotard in the context of Freudian *Nachträglichkeit* (Lyotard, *Heidegger and 'the jews,'* 59).

16 Heidegger, *Basic Questions of Philosophy: Selected "Problems" of "Logic"*, translated by Richard Rojcewicz and André Schuwer (Albany: SUNY Press, 1994), 89–90.

17 Here Heidegger inquires beyond Dasein when he talks about history; his question is what truth is today. This is already an epochal question – insofar as it contrasts today to other times. Historicity is not, as in the earlier essays, a matter of Dasein's being-in-the-world.

18 Heidegger, "Essence of Truth," 147. The essay hints at the historical significance of the essence of truth, e.g. with reference to "the liberation that grounds history"; "historical humanity"; and "the [] period in which the beginning of philosophy takes place" (Heidegger, "Essence of Truth," 152, 151, 152). *Basic Questions of Philosophy*, on the other hand, turns from the question of the essence of truth to a lengthy inquiry regarding the difference between historiography and history.

the primacy of the original temporal disclosure/concealment dynamic comes to the fore. That clue may lie in the very grammar of philosophy, for instance in Heidegger's treatment of the copula.[19] Perhaps the most compact example of Heidegger following such a grammatical clue is his 1955–1966 lecture course that was published as *The Principle of Reason*. Over several lectures, Heidegger ponders Leibniz's dictum "Nothing is without a reason [*nihil est sine ratione*]," and discovers a tonality for that dictum in which not the "*nihil*" but instead the "is" is emphasized. The second tonality indicates that the principle of reason can be interpreted as a statement not about entities but about being itself.[20]

Heidegger's historical narrative of the forgetting of being, following clues discovered in each epoch, looks to an ancient understanding that existed in an "originary" epoch. The anamnestic steps of Heidegger's investigations in the lectures are likewise translated into epochal stages of the history of being. Hence John Caputo argues that Heidegger transfers the event of unconcealment, which for Heidegger constitutes primordial truth, to a historical arc in which the understanding of the pre-Socratics was diminished in Aristotle and Plato and lost to the history of philosophy thereafter.[21] My own claim is as follows: with the history of metaphysics Heidegger makes a phenomenological investigation on the one hand and tells a historical/genealogical story on the other hand. Whereas the phenomenological inquiry looks, in quasi-transcendental style, to an event-like condition of possibility that is obscured by the present phenomenon or concept, the "historical" story of Greek *aletheia* phenomenalizes and historicizes that phenomenological-transcendental procedure. The historical inquiry goes "back" in time to a forgotten origin, whereas the phenomenological inquiry goes "underneath" a series of misconceptions to uncover a disclosive dynamic. In other words, Heidegger's history of metaphysics in effect combines an epochal history with an anamnestic phenomenology. The *history* of metaphysics

---

19 See for instance Heidegger, *Basic Problems of Phenomenology*, trans. Albert Hofstadter (Bloomington: Indiana University Press, 1988), 178–212.

20 Heidegger, *The Principle of Reason*, trans. Reginald Lilly (Bloomington: Indiana University Press, 1996).

21 John D. Caputo, "Demythologizing Heidegger: "Alētheia" and the History of Being," *The Review of Metaphysics* 41, no. 3 (1988), 519–546. See also David Farrell Krell, "On the Manifold Meaning of Aletheia: Brentano, Aristotle, Heidegger," *Research in Phenomenology* 5 (1975): 77–94; Heribert Boeder, "Heidegger's Legacy: On the Distinction of 'Aletheia'," trans. Marcus Brainard, *Research in Phenomenology* 28, (1998): 195–210; Jean Grondin, "L'aletheia entre Platon et Heidegger," *Revue de Metaphysique et de Morale* 87, (1982): 551–556; Mark A. Wrathall, *Heidegger and Unconcealment: Truth, Language, and History* (Cambridge: Cambridge University Press, 2011); and Charles R. Bambach, *Heidegger's Roots: Nietzsche, National Socialism, and the Greeks* (Ithaca: Cornell University Press, 2003), 189–194.

allegorizes in the narrative the anamnestic *procedure* of his phenomenological accounts, each of which uncovers an obscured essence, truth, or enactment that constitutes a dynamic of disclosure and concealment.

This comparison of the inquiry into the historicity of Dasein and the later history of being is not aimed at revisiting intra-Heideggerian debates on the *Kehre*.[22] Whether or not we see the *content* of Heidegger's interests as shifting from an earlier interest in human Dasein to a later interest in the history of being, the point is that the epochal story rehashes in another *form*, i.e., in the form of a historicizing narrative, the structure of the phenomenological approach that discovers the eventlike essence, truth, or "ground" of a concept or phenomenon.

The disappearance of history

If Heidegger's history of being allegorizes in narrative form the stepwise procedure of the investigation into the historicity of Dasein, this means that there is a considerable formal continuity in Heidegger's corpus, despite disagreement among scholars as to the continuity of Heidegger's thought and method across his career. The theme of continuity in Heidegger's corpus is fraught, above all because it raises the question of how important a role National Socialism played in that thought. Jürgen Habermas, to take one of the most significant respondents to this question, has emphasized the socio-political import of the transformation of Heidegger's notion of history between his earlier and later work: in earlier Heidegger, historicity is individualized and pertains to the possibility of Dasein, whereas in the later work historicity becomes generalized and is said to pertain to a *Volk*, the west, or the planet.[23] For Habermas this alteration evacuates Heidegger's later work of direct social relevance. Well before the publication of the Black Notebooks, Habermas writes that Heidegger "paid for the wealth of his later insights [...] with

---

22 Useful secondaries to this topic include: Orlando Pugliese, *Vermittlung und Kehre: Grundzüge des Geschichtsdenkens bei Martin Heidegger* (Freiburg: Alber, 1965); Laurence Paul Hemming, "Speaking Out of Turn: Martin Heidegger and die Kehre," *International Journal of Philosophical Studies* 6, no. 3 (1998): 393–423, 46; Henning Trüper, "Löwith, Löwith's Heidegger, and the Unity of History," *History and Theory* 53 (2014): 45–68.

23 Richard J. Bernstein, *The New Constellation: The Ethical-Political Horizons of Modernity/Postmodernity* (Cambridge: Polity Press, 1991), 81, demonstrates that "we can even interpret the twists and turns in his thinking – from at least 1929 on – as *in part* a response to his understanding of the political events he was witnessing."

a narrowing of his view to the dimension of a resolutely stylized history of metaphysics."[24] Worse yet, Heidegger's move away from Dasein opens his thought to ill-fated, happenstance sociopolitical perspectives and narratives, in other words, to National Socialism:

> Diese Abstraktion von gesellschaftlichen Lebenszusammenhängen hatte Folgen für Heideggers [...] Zugriff auf kursierende Zeitdeutungen. Je mehr die reale Geschichte hinter der "Geschichtlichkeit" verschwand, um so leichter konnte sich Heidegger auf einen naïf-prätentiösen Gebrauch von ad hoc aufgegriffenen Gegenwartsdiagnosen einlassen.
>
> This abstraction from the contexts of social life may be one reason for Heidegger's reliance on whatever interpretations of the age happened by. [...] The more real history disappeared behind Heideggerian 'historicity', the easier it was for Heidegger to adopt a naive, yet pretentious, appeal to 'diagnoses of the present' taken up ad hoc.[25]

Let us note in passing that Habermas's description condems Heidegger's uptake into his philosophy of certain elements that belong to, or at least are compatible with, National Socialism. But for the present argument, Habermas's attempts to explain precisely how that uptake came to pass are more significant. Habermas claims that Heidegger's intellectual withdrawal from the questions of Dasein's individual engagement with the social world accounts for that uptake, yet Habermas's phrasing does not lay upon Heidegger the full responsibility for "rel[ying] on" and "adopt[ing]" the interpretations and diagnoses of his time.

The withdrawal in Heidegger's work from the social context of Dasein and the seemingly agent-less "disappear[ance]" of "real history [...] behind Heideggerian 'historicity'" provide, for Habermas, "one reason" that made it "easier" for Heidegger to accept the historical interpretations supplied by National Socialism. Habermas portrays Heidegger in a passive role, as merely prone to adopt the interpretations of his day owing to real history's "disappear[ance]" and "abstraction." Habermas does not point out that Heidegger's own writing *produced* that disappearance of "real history" in his work, owing to its reformulation of historicity, which in turn made a ready place for those "naïve, yet pretentious," and ultimately genocidal, interpretations. Even more important, Habermas takes Heidegger's continued use, even in his very late work, of the language of history and historicity as a strategic maneuver to conceal the

---

[24] Jürgen Habermas, "Work and Weltanschauung: The Heidegger Controversy from a German Perspective," in *The New Conservatism: Cultural Criticism and the Historians' Debate*, ed. and trans. Shierry Weber Nicholsen (Cambridge: MIT Press, 1989), 143.
[25] Jürgen Habermas, "Martin Heidegger: Werk und Weltanschauung," in *Texte und Kontexte* (Frankfurt am Main: Suhrkamp, 1991), 52; Habermas, "Work and Weltanschauung," 143.

difference between the earlier phenomenology of Dasein and the later destinal and nationalist thematic. This is already apparent in Heidegger's 1947 "Letter on Humanism," as Habermas explains:

> Mit Hilfe einer Operation, die man "Abstraktion durch Verwesentlichung" nennen könnte, gelingt so die Entkoppelung der Seinsgeschichte vom politisch-historischen Geschehen. Diese wiederum erlaubt eine bemerkenswerte Selbststilisierung der eigenen philosophischen Entwicklung. Heidegger betont von nun an die Kontinuität seiner Fragestellung und bemüht sich, das Konzept der Seinsgeschichte durch Rückprojektion auf das unvollendet gebliebene Werk *Sein und Zeit* von den verräterischen weltanschaulichen Elementen zu reinigen.
>
> With the help of an operation that we might call "abstraction via essentialization," the history of Being is thus disconnected from political and historical events. This, again, allows for a remarkable self-stylization by Heidegger of his own philosophical development. From now on he emphasizes the continuity of his problematic and takes care to cleanse his concept of the history of Being from telltale ideological elements by projecting it back onto the never-completed *Being and Time*.[26]

Heidegger's consistent use of the terminology of history serves, in Habermas's view, to cover up a profound shift in Heidegger's own thought, to conceal a rupture in his thought, in which at least the rhetoric, and at most the actual doctrine, of National Socialism was incorporated into that thought.

Habermas's thesis regarding Heidegger's turn away from the phenomenology of Dasein to claims about western civilization suggests that Heidegger uses the language of history to produce an *appearance* of continuity between his earlier and later work. He implies that Heidegger did so in order to downplay the connection of his earlier work, as attested in the rhetoric of historicity, to National Socialism. This is a convincing claim for Heidegger's own interest in emphasizing the continuity within his own corpus. For the purposes of the present chapter, an intratextual or intra-corporal continuity is at issue: the argumentative emplotment, or formal procedure, that we have observed in the inquiries into historicity, art, and truth is translated into the epochal narrative of the history of being. In this way, Heidegger allegorizes the arc of inquiry within a historical narrative into a forgotten event-*Abgrund* that we have seen in his essays on historicity, art, and truth. He defines an anamnestic course, across epochs, that recapitulates the anamnestic procedure of his earlier inquiries into historicity, art, and truth. The history of metaphysics translates into story form the argumentative "plot" that we have described as structuring his phenomenological inquiry.

---

26 Habermas, "Werk und Weltanschauung," 71; "Work and Weltanschauung," 159.

Freedom and the history of being

The narrative of the history of metaphysics recasts as an epochal emplotment what has here been called the anamnestic procedure of Heidegger's earlier inquiries into concepts and phenomena – paradigmatically art, truth, and history. He finds in an epoch, that of the pre-Socratics, an understanding of truth and this epochal claim historicizes – i.e., narrate in terms of the past – the procedure that his phenomenological inquiry follows in examining the historicity of Dasein. That is, Heidegger locates in a specific period of history a knowledge-experience of disclosure, whereas his phenomenological inquiries discovered that an event of disclosure constitutes a quasi-essence, truth, ground, or condition of concepts and phenomena including truth and art. The stepwise procedure in those phenomenological investigations is narrativized and divided into epochs of knowledge. Hence a formal continuity connects the inquiries into the historicity of Dasein and the history of metaphysics.

The late Hayden White defines "narrativity" in terms of the evasion and occlusion in a historical account of the conditions, choices, and purposes associated with its construction.[27] Narrativity thus also masks the principles of emplotment that grant a historical account its unity and continuity. White argues that moralizing is inherent in narrativity. He asks whether narrative as such is possible to think without the moralizing effected by the very fact of narrative ending and closure, which gratify our profound desires for completeness and meaning. White's analysis of narrativity and how it occludes the strategic uses of a narrative provides crucial resources to ideology critique in historiography. Let us observe how those resources apply to the Heideggerian history of metaphysics: Heidegger represents the history of metaphysics as a unitary trajectory of the west, governed by its own law. In the 1937–1938 lectures he even defines history as the unfolding over actual historical time of an "innermost hidden law of the beginning."[28] This is quintessentially "narrativistic" in White's sense, for it represents as intrinsic to the events and epochs themselves an autonomous unfolding and periodization that are in fact produced, selected, and assembled by Heidegger himself. Indeed, critics have complained that Heidegger characteristically portrays his

---

[27] Hayden White, "The Value of Narrativity in the Representation of Reality," in *The Content of the Form: Narrative Discourse and Historical Representation* (Baltimore: The Johns Hopkins University Press, 1987), 1–25.

[28] Heidegger, *Basic Questions of Philosophy*, 35.

own inquiry as simply following free thinking, where in fact he steers it, sometimes awkwardly and even violently.²⁹

The evocation of the pre-Socratics as the proper pre-history of metaphysics to which we in the age of technology must return is a key "moralizing" component in White's sense. The "hidden law of the beginning" among the pre-Socratics bears hallmarks of Whitean narrativity. That reference ties beginning and end, unifies the account of the epochs of western civilization, and therefore provides the "moral" of Heidegger's epochal story. While White combines literary and ideology-critical insights in his description of narrativity as a feature of modern historiography, the same features can be found in Heidegger's approach to the history of philosophy. For as the 1937–1938 lecture course shows, he cites the knowledge-experience of the ancient Greeks as the "event" that constitutes the beginning of western philosophy and from which it subsequently falls, turns away, or simply forgets, and to which thinking should return – in fact, to which Heidegger's own philosophizing returns us.

That epochal story, as we have shown, historicizes the procedure of Heidegger's own phenomenological inquiry, namely in beginning with an extant phenomenon or concept and uncovering in stepwise fashion the obscured event of disclosure/concealment that forms its quasi-essence, Abgrund, truth, or condition. In historicizing across epochs the steps of his own earlier phenomenological procedure, Heidegger introduces a moralizing narrativity. He presents the history of western civilization as governed by a law, which unifies it and thereby forecloses other possibilities. While Dasein was defined in terms of possibilities and being-in-the-world, the "hidden law of the beginning" introduces narrativity into the history of being and therewith restricts possibility.

For Heidegger, the history of being therefore sounds a deterministic note. This is in contrast to the historicity of Dasein, which was defined as openness and possibility. The narrative of epochality, because it is emplotted as following a "law of the beginning," constrains the openness of Dasein's historicity. Because narratives incorporate plot, they *constrain* the very possibilities they represent. The anamnestic approach to Dasein's historicity, the origin of the work of art, or the essence of truth, on the other hand, emphasized the *openness* of the phenomenon or concept. The "emplotment" of Heidegger's earlier

---

29 Dieter Henrich claims that Heidegger's reading of Hölderlin's "Andenken" is violent, see Dieter Henrich, *The Course of Remembrance and Other Essays on Hölderlin*, ed. Eckart Förster, trans. Taylor Carman (Stanford: Stanford University Press, 1997), 294n94. See also Xiaomang Deng, "Heidegger's Distortion of Dialectics in *Hegel's Concept of Experience*," *Frontiers of Philosophy in China* 4, no. 2 (2009): 294–307. This is a central topic for critics of Heidegger and is treated widely.

inquiries into truth, art, and history results in a turn in each case to a dynamic of disclosure and concealment that bears no epochal implications and hence makes no claims on actual history. Insofar as Heidegger renders historicity as an epochal plot rather than a quality of Dasein, he produces an ambivalent terrain between phenomenological investigation and historiographical claim.

The imperfect overlap between phenomenological and historical inquiry appears also in Hannah Arendt's political phenomenology. Both Heidegger and Arendt turn, albeit in different ways, to narrative and history when they seek to illuminate possibility, freedom, and spontaneity – each of which is not perfectly suited to the tight connection of events entailed by narrative. As we will see in the next chapter, Arendt turns to history, in particular the neoteric history of totalitarianism, in order to investigate the shape of human freedom in a shared world. For Arendt, the phenomenon of freedom as spontaneity and the possibility of newness is at the heart of her historical investigation of totalitarianism. Arendt's account of the destruction of freedom under totalitarianism teeters between contingency and plotlike necessity: she looks for precedents and ingredients to make for historical continuity to her phenomenology of totalitarianism, even where she insists on the unprecedented nature of totalitarianism and the contingency of history in each case.

The function of historical narrative in Arendt's account, as we shall see, is in a way the converse of its function in Heidegger. For Heidegger the historical narrative allegorizes the procedure of phenomenological investigation, locating the experiential event of disclosure and concealment in a distant epoch and implicitly constraining human history. For Arendt a narrative of the recent totalitarian past provides the ultimate illumination of the quasi-structural condition of human freedom. The historical phenomenology she conducts in *Origins of Totalitarianism* uses narrative in order to uncover the conditions and possibilities of human freedom. As we will show, however, the tight connection of events in narrative interferes with the representation of human free spontaneity in which she is profoundly invested. We have seen that Heidegger's history of being seems to foreclose spontaneity, connecting the epochs with the law of the beginning. Arendt's treatment of totalitarianism seeks to spotlight the nature of human freedom and the structural conditions that it requires; it makes use of storytelling in order to perform a phenomenology of possibility, spontaneity, and newness. Heidegger's history of being is phenomenology that becomes storytelling, an embedding of a form of inquiry into an emplotment with a moralizing and narrativizing "law" that unifies beginning and end, which thus appears to foreclose the possibility, spontaneity, and newness of an open future.

7 Arendt's Epochal Phenomenology: History and the New

In *Origins of Totalitarianism* Arendt undertakes a phenomenological, even what we might call structural, investigation into what she claims are utterly new circumstances in the history of humankind. Totalitarianism constitutes, Arendt writes, a "novel form of government" that "differs essentially" from all its predecessors.[1] It produced a "new kind of human being" and "developed entirely new political institutions."[2] The book scrutinizes in historical detail this novel form of government, new kind of human being, and new political institutions, along with what Arendt characterizes as the new phenomena of anti-Semitism and nineteenth-century European imperialism.[3] *Origins of Totalitarianism* also analyzes the centrality of newness, in the form of spontaneity, to human freedom and politics. Arendt portrays totalitarianism as threatening newness by "forc[ing] [mankind] into that gigantic movement of History or Nature" (*OT* 473). She argues that the logic of totalitarianism, embodied in its institutions, models human activity as a part of an inexorable historical process. That model of inexorable process threatens to destroy the spontaneity of human freedom. The effects of the new phenomenon of totalitarianism in turn illustrate the need for a "new political principle" (*OT*, ix).[4]

The emphasis on newness is symptomatic of Arendt's larger interest, throughout *Origins of Totalitarianism* and her entire corpus, in the phenomena of human spontaneity, action, and freedom. As we have seen in Chapter Four, for Kant freedom is a spontaneous and hence non-natural causality. Arendt's

---

[1] Hannah Arendt, *Origins of Totalitarianism* (New York: Harcourt, 1976), 460, hereafter cited in text as *OT*.
[2] First reference is to Elisabeth Young-Bruehl, *Hannah Arendt: For Love of the World* (New Haven: Yale University Press, 1982), 152; second reference is to *OT*, 460.
[3] See also William Selinger, "The Politics of Arendtian Historiography: European Federation and *The Origins of Totalitarianism*," *Modern Intellectual History* 13, no. 2 (2016): 417–446; Zenonas Norkus, "Why Hannah Arendt's Ideas on Totalitarianism are Heterodox," *Topos: Journal for Philosophical and Cultural Studies* 2, no. 19 (2008): 114–136; Wayne Allen, "Hannah Arendt and the Ideological Structure of Totalitarianism," *Man and World: An International Philosophical Review* 26, no. 2 (1993): 115–129.
[4] See also Michael H. McCarthy, *The Political Humanism of Hannah Arendt* (Lanham MD: Lexington Books, 2014); Annette Vowinckel, *Arendt* (Leipzig: Reclam, 2006), 23–40; Hauke Brunkhorst, "*The Origins of Totalitarianism*: Elemente und Ursprünge totaler Herrschaft," in *Arendt-Handbuch: Leben, Werk, Wirkung*, eds. Wolfgang Heuer, Bernd Heiter and Stefanie Rosenmüller (Stuttgart: J.B. Metzler, 2011), 35–42.

historical phenomenology is indebted to the Kantian formulation, for instance when she defines "[s]pontaneity [as] man's power to begin something new out of his own reactions" (*OT*, 455) and also characterizes "[f]reedom as an inner capacity of man [that] is identical with the capacity to begin" (*OT*, 473). Arendt's work investigates how free will constitutes, in the Kantian sense, a perduring human possibility and the *sine qua non* of politics, where politics is defined as a sphere of free human action. Hence Arendt's 1963 *On Revolution*, for instance, interrogates the manifestation of human freedom in concrete historical moments.[5] Nonetheless, there are significant differences between Kant's treatment of the role of freedom in the representation of universal history and Arendt's use of historical narrative to depict the modern epoch and what she considers the worldlessness of totalitarianism. For Kant we can "see" free will in history only in the context of the well-known "as-if," namely if we regard history in accord with the idea of progress, as described in Chapter Four of this book. Arendt is not preoccupied with the Kantian "as if" that allows us subjectively to *view* history *as* both free and governed by reason. She seeks rather to understand *actual* history in *Origins of Totalitarianism*. Arendt undertakes to expose the *real* operation of human freedom in recent history and the *real* threat posed to it by totalitarianism. Her historical narratives leap over the "as-if" in the Kantian account of our knowledge of the world. They offer a historical phenomenology of freedom, its manifestations, and its eclipse, rather than a transcendental account of the possibility of freedom. In other words, the *representation* of history is at stake with Kant; Arendt's "realist" treatment of history is, in contrast, a matter of the *actual* eruption of human freedom in its singular appearances on the scene of history.

In its specificity and concreteness Arendt's recourse to history appears to be more conventional than Heidegger's history of metaphysics, which was examined in the previous chapter. She writes about actual events and figures of the past, for instance Rosa Luxemburg, the French Revolution, and the development of atomic weapons.[6] She describes history in terms of epochs, for instance in *The Human Condition*'s account of the rise of the social in modernity and the corresponding decline of the political.[7] Nonetheless, while Arendt treats the epochs of history more concretely than does Heidegger – and with more actual historical

---

[5] Arendt, *On Revolution* (New York: Penguin, 1963).
[6] Arendt, *Rahel Varnhagen: The Life of a Jewess* (Baltimore: The Johns Hopkins University Press, 1997); "Europe and the Atomic Bomb," in *Essays in Understanding 1930–1954: Formation, Exile and Totalitarianism*, ed. Jerome Kohn (New York: Harcourt, Brace & Co., 1994), 418–422.
[7] Arendt, *The Human Condition* (Chicago: University of Chicago Press, 1998).

data – her primary objective is not historical. For ultimately Arendt is in pursuit of a phenomenology of newness and beginnings, and hence of discontinuity, rather than a historical account per se. Newness implies discontinuity with what came before; the centrality of newness and discontinuity makes Arendt's historical phenomenology entirely different from Kant's universal history and Heidegger's history of metaphysics, each of which relies upon a "red thread" or through-line in order to constitute a linear narrative.[8]

What is significant for the present argument is that the "real" history that is recounted in *Origins of Totalitarianism* inadvertently evokes a tension between the narrative aspects of historical description and the phenomenology of newness. This chapter will argue that the requirements of historical narrative, and above all the requirement for one-by-one connection, are not entirely compatible with Arendt's phenomenological inquiry into human freedom. Phenomenology on the Husserlian model, to which Arendt and Heidegger are indebted, tends toward the isolation of a static phenomenon – as in Husserl's proverbial semester-long seminar spent analyzing a mailbox.[9] That essentializing moment in phenomenology is a forerunner also of structuralist literary criticism by way, in large part, of Roman Jakobson and Roman Ingarden.[10] When recounting historical events in which human freedom is said to come to light, however, Arendt

---

[8] Arendt's historical phenomenology diverges in other ways from Heidegger's epochal narrative. As we have seen in the previous chapter, Heidegger's history of metaphysics takes him away from the early analysis of *Mitsein*; Arendt's corpus, in contrast, can be seen as an investigation of how we live in a common world with others. This "love of the world," or *amor mundi*, motivates her inquiries into singular historical moments, e.g. the French revolution, totalitarianism, and the U.S. civil rights movement. Arendt interrogates the centrality of *Mitsein*, a task that, to be sure, takes her into empirical moments where historical events come to pass. That historical inquiry is conducted in the service of a phenomenological inquiry into newness and freedom in the shared human world. For Arendt the epochal story of totalitarianism hopes to further the investigation and worldly restoration of the human freedom that is her profound concern.
[9] Martin Heidegger, *Ontology: The Hermeneutics of Facticity*, trans. John van Buren (Bloomington: Indiana University Press, 1999), 86.
[10] See Roman Jakobsen, "Zur Struktur des Phonems," in *Selected Writings I: Phonological Studies* (Berlin: De Gruyter, 2012), 280–310; Beate Stawarska, *Saussure's Philosophy of Language as Phenomenology: Undoing the Doctrine of the Course in General Linguistics* (Oxford: Oxford University Press, 2015); Jonathan Culler, "Phenomenology and Structuralism," *Human Context* 5 (1973): 35–41; Hugh Silverman, *Inscriptions: Between Phenomenology and Structuralism* (New York: Routledge, 1987); Elmar Holenstein, "Jakobson and Husserl: A Contribution to the Genealogy of Structuralism," *Human Context* 7 (1975): 61–83; Roman Ingarden, *The Literary Work of Art* (Evanston: Northwestern University Press, 1973); Hans H. Rudnick, "Roman Ingarden's Literary Theory," *Analecta Husserliana* 4 (1976): 105–119.

incorporates narrative connections in order to produce a properly historical account. That connective requirement of conventional historical narrative interferes or competes with the phenomenological analysis of the spontaneity of human freedom.

This chapter will illuminate this interference specifically where Arendt insists on the newness of totalitarianism and anti-Semitism and yet makes use of historical material in such a way as to inadvertently import continuity. Whether Arendt's accounts of anti-Semitism and totalitarianism are historically accurate or even defensible is not the issue here; what is at issue is how elements of historical narrative, problematic or not regarding their accuracy, contravene her focus on newness. To begin with, Arendt's foray into the history of anti-Semitism evokes a narrative continuity, although she proposes to examine anti-Semitism as an entirely new phenomenon. The synthetic operations of narrative trammel her phenomenological goal of explaining anti-Semitism and totalitarianism as utterly new. As we will see, examining a small portion of *Origins of Totalitarianism* – and also Arendt's essays "The Concept of History: Ancient and Modern" and "What is Freedom?" – the reflection on newness comes into conflict with a historicizing phenomenology.

Anti-Semitism in *Origins of Totalitarianism*

Within her mammoth study of totalitarianism, Arendt's controversial treatment of anti-Semitism reflects an underlying tension between a phenomenology of newness and a historical account with narrative continuity. Above all she condemns the illusion of continuity perpetuated by theories of eternal anti-Semitism, insofar as they obscure, in her account, the unprecedented character of modern anti-Semitism.[11] Newness is here central. Her 1967 preface to the book's section on anti-Semitism emphasizes the difference between the phenomena of medieval Jew-hatred and modern anti-Semitism: "Anti-Semitism [...] and religious Jew-hatred [...] are obviously not the same; and even the

---

[11] There are numerous debates and critiques surrounding Arendt's account of the history of anti-Semitism. See for instance Peter Staudenmaier, "Hannah Arendt's Analysis of Antisemitism in *The Origins of Totalitarianism:* A Critical Appraisal," in *Patterns of Prejudice* 46, no. 2 (2012): 154–179; Michele Battini, "The Birth of an Anti-Jewish Anti-Capitalism," *Constellations: An International Journal of Critical and Democratic Theory* 16, no. 4 (2009): 615–633; Natan Sznaider, "Hannah Arendt and the Sociology of Antisemitism," *Österreichische Zeitschrift für Politikwissenschaft* 39, no. 4 (2010): 421–434; Richard J. Bernstein, *Hannah Arendt and the Jewish Question* (Cambridge: MIT Press, 1996).

extent to which the former derives its arguments and emotional appeal from the latter is open to question" (*OT*, xi). Arendt first explains this difference in terms of a discontinuity of epochs, calling "fallacious" any "notion of an unbroken continuity of persecutions" (*OT*, xi). Arendt argues in a compact way for the existence of breaks and gaps in the history of Jewish-Christian relations. Her argument relies on a 1962 publication by historian Jacob Katz, who claimed that a hiatus in Christian persecution of Jews in Europe occurred from the sixteenth to eighteenth centuries. Katz provides historiographical support for Arendt's rejection of a mythos of a continuous history of Christian persecution of Jews.

While Arendt relies on Katz's book in her depiction of an epochal difference between medieval Jew-hatred and modern anti-Semitism, she also indirectly asserts the existence of a continuity between them. During the hiatus in persecution, according to Arendt's citation of Katz, Jews themselves allegedly formulated the idea of an inner essential difference between Jews and Christians. Katz refers at some length to Maharal, or Judah Loew ben Bezalel of Prague, who maintained that Jews possess a special nature and for that reason were chosen by God.[12] In Arendt's rendition of Katz, this formulation of an essential Jewish difference, which originated among Jews themselves, was taken over only later by Christians. That older Jewish notion of an essential Jewish difference became a "condition *sine qua non* for the birth of anti-Semitism" (*OT*, xii).

It is beyond the purview of this book to consider whether or not Katz's historical narrative is accurate, i.e., when or whether Jews themselves originated a notion of an essential Jewish difference, or whether the theology of biblical Judaism provided the basis for Maharal's formulation. What is significant for this chapter's argument is that Arendt's recourse to Katz unsettles her central claim about the *absolute* newness of modern anti-Semitism. In Arendt's appropriation of Katz's account, an avowal of Jewish difference, formulated by Jews well before the nineteenth century, was taken over by Christians and thereby became available for modern anti-Semitism. Arendt's narrative of the origin and career of the notion of an essential Jewish difference thereby sets up a continuity between the Jewish identity formation of the early modern epoch and the racialism of the twentieth century. In other words, Arendt's argument for discontinuity in the history of Jewish-Christian relations reintroduces a narrative continuity to that history; she insists that modern anti-Semitism is entirely different from medieval Jew-hatred, *but also* uses Katz's historical account in order to intimate that modern anti-Semitism has roots in, and hence

---

12 Jacob Katz, *Exclusiveness and Tolerance: Studies in Jewish-Gentile Relations in Medieval and Modern Times* (London: Oxford University Press, 1961), 140.

some continuity with, a supposedly Jewish notion from the pre-Enlightenment epoch. Arendt borrows from Katz in order to support her argument that there was a discontinuity wherein something truly new took place with the Jewish formulation of an essential Jewish difference; this formulation marks an implicit epochal shift whose effects only appeared later with modern anti-Semitism. But then how completely new is anti-Semitism if its "condition *sine qua non*" emerged hundreds of years earlier? The continuity that Arendt introduces into her historical narrative undercuts the assertion of the newness of anti-Semitism in the modern epoch.

Let us pause to acknowledge the controversial nature of Arendt's use of Katz and the connection it suggests between causality and responsibility. In defining a supposedly Jewish idea of an essential Jewish difference as a *sine qua non* for modern anti-Semitism, Arendt seems to connect the persecution of twentieth-century Jews to a formulation that supposedly originated with early modern Jews. The thesis that Jews themselves formulated the notion of an essential Jewish difference is controversial because it seems to ascribe some responsibility – an indeterminate degree, but incendiary in any case – for modern anti-Semitism to Jews themselves. In Arendt's appropriation of Katz's narrative, Jews created the formulation that became anti-Semitism's "condition *sine qua non*." The connection between a sixteenth-century formulation and a nineteenth- and twentieth-century anti-Semitism is a slippery one, as is the question of whether responsibility can be assigned based on such a slippery connection. The issue would be even more ambiguous if instead of the phrase "condition *sine qua non*" Arendt had used the word "ingredient." This was the solution used by the editor of *The Portable Arendt* in his introduction as a way to sidestep the thorny terminology of cause.[13] This spectrum of connection from "ingredient" to "cause" is controversial because the words evoke, to different degrees and in different ontological registers, the notion that Jews themselves bear some degree of responsibility for anti-Semitism. Indeed, Arendt's account of modern anti-Semitism can be read as fluctuating between proposing reasons, causes, and conditions for anti-Semitism, each of which implies a different kind and degree of continuity between epochs. The question of how *entirely*

---

[13] See Peter Baehr, introduction to *The Portable Hannah Arendt* (New York: Penguin, 2000), x: "Anti-Semitism was not inevitable, nor should it be thought of as a necessary, perverse ingredient in fashioning or maintaining Jewish solidarity"; and Margaret Canovan, *Hannah Arendt: A Reinterpretation of Her Political Thought* (Cambridge: Cambridge University Press, 1992), 51: "This really new and lethal ingredient [that differentiated National Socialism from nineteenth-century anti-Semitism] was the version of anti-Semitism developed under imperialist influence by the Pan- movements of eastern Europe."

new anti-Semitism is testifies to the problem of defining the continuity and discontinuity between Jew-hatred and modern anti-Semitism, between early modernity and the twentieth century, and between historical event and historical responsibility.[14]

Arendt's reference to Katz and the formulation of an essential Jewish difference in early modernity is but one small instance in *Origins of Totalitarianism* of her incendiary ambiguity around ascriptions of responsibility for modern anti-Semitism. Her more prominent and lengthy historical account of the role of Jews in the new European nation-states contains similar controversial claims around the spectrum of responsibility – from cause to ingredient – insofar as she attempts to argue that the fate of European Jews was neither accidental nor random. Arendt famously characterizes European Jews as politically naïve, insular, and as involved, owing to their riches, in state finance, in order to explain the origins of modern anti-Semitism (*OT*, 11–53). In *Anti-Judaism: The Western Tradition*, David Nirenberg explains Arendt's use of such anti-Semitic tropes. Nirenberg's explanation indirectly illuminates Arendt's equivocation around the terminology of connection, ingredient, condition, and cause when describing the relationship between Jewish actions and anti-Semitism.[15] Nirenberg portrays Arendt as seeking to understand the "cultural sense" that made modern anti-Semitism convincing and effective. He argues that in the course of that seeking Arendt drew upon and participated in a longstanding western confusion between figural or "imaginary" Jews and actual Jews (Nirenberg, 462–463). Arendt's error, he writes, was to believe that anti-Semitic ideology "made cultural sense because it described something that the Jews really were, something that they really did" (Nirenberg, 463). In other words, Arendt mistook "imaginary" Jews for real Jews – e.g., in her hackneyed criticisms of nineteenth-century Jewish bankers and in her overstatement of the significance of Jews to bourgeois capitalism (Nirenberg, 463).

---

14 It is worth noting in this context that even Arendt's own reference to Katz is framed in epochal terms. She reflects briefly on Katz's significance for Jewish historiography and paints him as part of a completely new phase in an old endeavor. In her footnote to Katz Arendt heralds his work as the end of an epoch in Jewish historiography: "With [a younger generation of Jewish historians], the lachrymose presentation of Jewish history [...] has indeed come to an end" (*OT*, xii). Thus even the source text for her claim about epochal shifts in Jewish-Christian relations is deemed epoch-making in its own academic sphere, marking the end of a stage in the history of Jewish history. Although he provides the ambiguous claims for both a hiatus in Jewish-Christian relations and for the connection between a Jewish formulation and later Christian anti-Semitism, Katz himself is represented as part of yet another new phenomenon, namely the post-lachrymose Jewish historian.
15 David Nirenberg, *Anti-Judaism: The Western Tradition* (New York: Norton, 2013).

The notion of the "cultural sense" of imaginary vs. real Jews and what kind of responsibility real Jews bear for anti-Semitism also applies more narrowly to Arendt's epochal account of anti-Semitism. In Nirenberg's argument, Arendt explained the new phenomenon of anti-Semitism on the basis of a Jewish history of her own time that rested on anti-Semitic tropes. She accounts for the appearance of a novel anti-Semitism, supposedly disconnected from medieval Jew-hatred, on the basis of tropes and traditions of anti-Semitism itself. Nirenberg makes a case that Arendt's "facts" (Nirenberg's scare quotes) about the Jews, which lead her to wavering ascriptions of responsibility for anti-Semitism to Jews themselves, made such cultural sense because those "facts" fit with and appealed to longstanding "habits of thought" (Nirenberg, 462). Arendt's version of Jewish history therefore arose from an unthematized historical trope. Her "facts" made use of "critical concepts [...] [that were] themselves produced by a history of criticizing Judaism" (Nirenberg, 465). In other words, Arendt unwittingly drew on elements of Jew-hatred in formulating her understanding of why anti-Semitism was so powerful and attractive, mistaking those elements for actual historical facts. The irony is that Arendt's historical phenomenology casts about for real reasons, conditions, origins, or acts that would help account for anti-Semitism. She "bets on reality," to use a term from Rei Terada, without recognizing that the reasons, conditions, and origins she recounts are in fact non-originary, i.e., they are tropes and imaginings that have acquired their worldly reality and "cultural sense" simply in their repetition.[16]

The central issue for this chapter, however, is that on the one hand Arendt wants to say that modern anti-Semitism was new and unprecedented, while on the other hand, she explores its "conditions" – which constitute neither ingredients, predecessors, reasons, or causes. Nonetheless, her explorations of those "conditions" of anti-Semitism narrate a continuity of an indeterminate degree with previous historical events. In a nutshell we see here the difficulties of making a historical claim for newness in the context of Arendt's phenomenological study: the sixteenth-century hiatus in Jewish-Christian relations is supposed to stand as the marker of a new Jewish self-perception and hence inaugurate a new epoch that forms the condition of the even newer phenomenon of modern anti-Semitism. Anti-Semitism's uptake of that supposed historical precedent would thereby seem to make it not a completely new phenomenon after all. The supposedly new sixteenth-century Jewish self-definition constitutes a "condition *sine qua non*" that precedes and conditions, but does not produce,

---

[16] Rei Terada, "Thinking for Oneself: Realism and Defiance in Arendt," in *ELH* 71, no. 4 (2004): 860.

modern anti-Semitism – hence the question of responsibility in the historical narrative. Arendt wants to emphasize the newness of anti-Semitism, but her historiographical representation of its "condition *sine qua non*" seems to elide that newness and locate in the sixteenth century the "origins" of what she claims to be a categorically new nineteenth-century anti-Semitism.

Freedom or narrative: Competing Kantianisms

While her claim for the newness of modern anti-Semitism forms the center of Part I of *Origins of Totalitarianism,* that book's larger argument concerns the absolute newness of totalitarianism in modernity. The very word "origins" in the book's title appears to signal a historical approach to totalitarianism, i.e., an account of how totalitarianism came to exist in history. In fact, the book's aim is to produce a *phenomenology* – and not actually a history – of totalitarianism. Correspondence in 1951 between Arendt and Eric Voegelin indicates that the title of the book posed a problem for Arendt herself:

> I did not write a history of totalitarianism but an analysis in terms of history. […] The book, therefore, does not really deal with the "origins" of totalitarianism – as its title unfortunately claims – but gives a historical account of the elements which crystallized into totalitarianism.[17]

Arendt here backs off from her own title because it promises an account of "origins" where what she intended was an account of "elements." This reflects an inadvertent interference of the demands of historical narrative in the phenomenological project. In the ellipses marked above, Arendt emphasizes the "elemental," and thus phenomenological, focus of the book's investigations:

> I did not write a history of anti-Semitism or of imperialism, but analyzed the element of Jew-hatred and the element of expansion insofar as these elements were still clearly visible and played a decisive role in the totalitarian *phenomenon* itself.[18]

From this it is clear that Arendt did not seek to produce a proper historical account of totalitarianism, but rather to understand its elements and their role in the "totalitarian phenomenon." The goal, in other words, was phenomenological rather than historical. The claim for the newness of totalitarianism is, on its face, a claim about history and yet it ultimately serves a phenomenological,

---

[17] Arendt, "A Reply to Eric Voegelin," in *The Portable Hannah Arendt*, 158.
[18] Arendt, "Reply to Eric Voegelin," 158, emphasis mine.

rather than historical, investigation. Indeed, *Origins of Totalitarianism* constitutes a historical phenomenology, a study of a phenomenon and its elements, as they appear in their particular historical context.

While such an "elemental" focus is apparent from the correspondence with Voegelin, Arendt's insistence on the lack of precedent for totalitarianism, which implies a historical discontinuity, also marks the book's divergence from conventional historiography. It is with an anti-historicizing thrust that Arendt writes in her preface to the first edition to *Origins of Totalitarianism* that her goal is to comprehend history, to interpret it, including "facing up to, and resisting [...] reality," without, as she emphasizes, "deducing the unprecedented from antecedents" (*OT*, viii). The task of the book is to see totalitarianism in its thoroughgoing discontinuity; the unprecedented must not be deduced from antecedents – in other words, there is no governing connection between preceding events and totalitarianism itself. The newness of totalitarianism means that it is not to be portrayed according to the Aristotelian requirements for tragic plot, in which intrinsic connection is required.

Newness is at stake at still other levels in Arendt's treatment of totalitarianism. As we have seen, she argues that the phenomenon of totalitarianism was unprecedented, something absolutely new, although it had precursors and conditions.[19] Particularly new, and especially pernicious, is totalitarianism's goal of *extinguishing* the newness that she identified as the defining criterion of human action: "Totalitarian domination [...] will eliminate precisely the capacity of man to act" (*OT*, 467). What is more, "[t]otal terror destroys" the condition of political freedom that she identifies as "the space between men as it is hedged in by laws" (*OT*, 466).[20] The book's examination of the elements of totalitarianism thus also encompasses a phenomenology of the human spontaneity and freedom that totalitarianism annihilates. As a counterpoint to the phenomenology of totalitarianism, the book's final chapter, an essay on ideology and terror, closes with an invocation of beginnings. Arendt quotes Augustine, who within the theological tradition connects the creation of human beings with the inauguration of beginnings. Beginning "is indeed every man," "is the supreme capacity of man," and

---

[19] On newness and the unprecedented, see also Nancy Fraser, "Hannah Arendt in the Twenty-First Century," *Contemporary Political Theory* 3, no. 3 (2004): 253–261; Mark Reinhardt, "What's New in Arendt?" *Political Theory: An International Journal of Political Philosophy* 31, no. 3 (2003): 443–460; Taran Kang, "Origin and Essence: The Problem of History in Hannah Arendt," *Journal of the History of Ideas* 74, no. 1 (2013): 139–160; and Linda Zerilli, "Castoriadis, Arendt, and the Problem of the New," *Constellations* 9, no. 4 (2002) 540–553.

[20] On terror see Dana Villa, "Political Violence and Terror: Arendtian Reflections," *Ethics and Global Politics* 1, no. 3 (2008): 97–113.

"politically, it is identical with man's freedom" (*OT*, 479). This is for Arendt a significant element of the freedom that totalitarianism destroys. Specifically, it is "the capacity to begin," understood "as an inner capacity of man" (*OT*, 473). Underscoring the human capacity to begin, Arendt emphasizes the unconditionality of the beginning:

> Over the beginning, no logic, no cogent deduction can have any power, because its chain presupposes, in the form of a premise, the beginning. [...] [W]ith the birth of each new human being a new beginning arise[s] and raise[s] its voice in the world.[21] (*OT*, 473)

For Arendt, totalitarianism, itself a novel phenomenon in the history of humankind, threatens to vitiate the human capacity for new beginnings.[22]

Arendt seeks to understand how totalitarianism came to exist, but also insists on its radical newness and disconnection from previous epochs. Perhaps this explains why she seems to resort to a spectrum of what we might call "soft connections." For instance: "The road to totalitarian domination leads through many intermediate stages for which we can find numerous analogies and precedents" (*OT*, 440). These analogies and precedents do not amount to the "hard connection" of causality, but they permit connections between "stages" to be perceived. As George Kateb writes about Arendt's book, "[T]he totalitarian movement [...] *wears the appearance* of earlier movements, racist and imperialist, in Europe."[23] Kateb's stress on appearance highlights the ambiguity of the connections Arendt wants to make between totalitarianism and other movements. Precisely what connection to the present is implied by "wearing the appearance" of a past movement – is it a superficial connection or a profound one? Arendt writes that causality is alien within historical science, that an event cannot be deduced from its own past. For this reason there is a radical contingency in history, because human events are not *necessarily* connected,

---

21 On natality in Arendt, see Margarete Durst, "Does the World Exist?" in *Analecta Husserliana: The Yearbook of Phenomenological Research* 79 (Dordrecht: Kluwer, 2004), 777–797; Mavis Louise Bliss, "Arendt and the Theological Significance of Natality," *Philosophy Compass* 7, no. 11 (2012): 762–771; Peg Birmingham, *Hannah Arendt and Human Rights: The Predicament of Common Responsibility* (Bloomington: Indiana University Press, 2006), 4–34.

22 The second form of freedom that totalitarianism destroys is indicated above in Arendt's reference to "the space between men as it is hedged in by laws." Here freedom takes on a decidedly spatial figuration; it is "freedom as a political reality [...] identical with a space of movement between men" (*OT*, 473). This element of freedom is not separate from the capacity to begin, but involves *Mitsein*, or coexistence.

23 George Kateb, *Hannah Arendt: Politics, Conscience, Evil* (Totowa, NJ: Rowman and Allanheld, 1983), 56, emphasis mine.

for at each turn human freedom remains open. Nonetheless, this exposes a conundrum for her own investigation, which seeks explanations for the appearance of the new phenomenon of totalitarianism and leads to claims for the kinds of "soft connections" of ingredient, precedent, analogy, and *sine qua non* that this chapter has described, i.e., connections less categorical than those defined by the terminology of "cause" and "origin."

The emphasis on newness and beginnings is clearly crucial to Arendt's concept of freedom and her explanations of the phenomenon of totalitarianism. Her historical material, however, entangles her phenomenological analyses of elements and conditions with diachronic, emplotted narratives. The emplotted character of her narratives in turn forecloses the absolute contingency that freedom seems to require.[24] In turning to the history of National Socialism and Stalinism, Arendt ends up with more narrativity and hence more plot – more connection of events – than her interest in freedom and spontaneity would seem able to bear. For there *are* moments in *Origins of Totalitarianism* that demonstrate a commitment to a continuous, even plotlike history. She writes in the preface to the first edition, for instance, that her goal is to uncover what she calls "the hidden mechanics" that dissolved the political and spiritual elements of our human world in totalitarianism (*OT*, viii). She also writes in that preface that "The subterranean stream of western history has finally come to the surface" (*OT*, ix). Both figurations, mechanistic and naturalistic, represent the events and situations to be recounted as insuperably connected. They hint at a "plotlike" underlying structure that has determined the events and phenomena that Arendt goes on to investigate. They intimate that there are inexorable forces at work in this history, forces which would seem to undermine her goal of affirming the phenomenon of human freedom.

This entanglement of emplotments and connections in Arendt's phenomenology of freedom is symptomatic once again of the contradiction that shapes epochal phenomenology and *The Origins of Totalitarianism*. Arendt strives to produce a phenomenology of totalitarianism and of its different elements and also to argue that this new phenomenon destroys worldly human freedom and spontaneity. She builds that argument on the basis of historical material and interpretation. But the recourse to historical material poses a dilemma as well. As Kant already made clear, to give a historical account a unity and coherence demands a narrative connection between events. Insofar as Arendt makes

---

24 One way to view this entanglement is as a clash between the Kantianism of White, who emphasizes the sheerly narrative synthesis of events, and the Kantianism of Arendt, who emphasizes new beginnings as freedom.

historical claims, as the word "origins" in the title suggests, the intrinsic emplottedness of the book's narrative interferes with the representation of totalitarianism as unprecedented. Arendt risks on the one hand giving an account of totalitarianism that appears either too deterministic – as if totalitarianism happened inevitably and was bound to be the way it was, as if it were the plot of modernity that drew the world into its story, or on the other hand, if totalitarianism were portrayed as entirely contingent, unprecedented, and disconnected from historical causality, it would be difficult to point in a phenomenological fashion to an essential character of totalitarianism and the freedom it destroys. This means that Arendt's account of totalitarianism risks being both too historical for her phenomenological goals and not phenomenological enough to fulfill the demands for emplotment in the historical account.

The totalitarianism of "process"

Arendt's rejection in her essays "The Concept of History: Ancient and Modern" and "What is Freedom?" of a conception of history as a "process" indicates the same concerns we have observed about freedom and its depiction in historiographical narrative. That is, she argues that any historiography that models history as process undermines the human possibility that corresponds to newness.[25] For the representation of history as a matter of processes destroys our own sense of human spontaneity and hence freedom. The figural reverberations of a language of process in historiography, in other words, has repercussions for worldly free action. Nonetheless, it is remarkable that Arendt looks at the rise of a logic of newness-destroying process in scientific modernity as itself a process of sorts. Her account runs as follows: in the development of the modern concept of history, and in accord with the emergence of modern science, history comes to be represented as a process. This marks an epochal shift in the thinking of history that risks losing sight of

---

[25] Arendt, "The Concept of History: Ancient and Modern," in *Between Past and Future: Eight Exercises in Political Thought* (New York: Penguin, 1968), 79. "Process" is crucial to ideology, as Arendt explains in *Origins of Totalitarianism*. She defines ideology as "the logic of an idea [...] [whose] subject matter is history" (*OT*, 469). Ideology, in other words, interprets the course of history as the unfolding of a particular logic or idea, rather than, for instance, as a contingent sequence of free human actions. Totalitarian ideology submits human beings to a logic of "process" that "dispenses with the human will to action altogether" (*OT*, 468). It even threatens the possibility of new thinking: "[T]he self-coercive force of logicality is mobilized lest anybody ever start thinking" (*OT*, 473).

freedom as spontaneous, non-predetermined action. For if human history is represented as a process, Arendt argues, then individual events, people, and freedom lose their unique significances:

> [T]he process of history [...] cannot bestow meaning on particular occurrences [...] because it has dissolved all of the particular into means whose meaningfulness ends the moment the end-product [of the process] is finished. ("History," 80)

Perhaps we here glimpse Arendt's debt to Benjamin's historical materialism, with its stress on the flashing up of singular moments, in contrast to triumphalist or progressivist narrativity.[26]

Arendt's "Concept of History: Ancient and Modern" considers the purposes that historiography has served and how these have changed, such that historiography itself has changed and become subject to the metaphorics of process. The essay describes historiography in modernity as compared to the greatness of Homerian and Herodotan history. For the latter, and for the ancient world in general, history was a matter of preserving greatness. Arendt writes, in an echo of Hegel, that ancient history undertook the remembrance of greatness in deeds and words.[27] What is more, the notion of historical "objectivity" in modern historiography marks a significant divergence from the ancient value of "impartiality": Homer showed impartiality in history, praising Greeks and Trojans alike. Likewise the corollary of objectivity was for the pre-Socratics a matter of recognizing different points of view on one and the same situation, i.e., a form of respect for doxa as one's individual viewpoint in a world of plural viewpoints ("Concept of History," 49–53).

With the modern understanding of history as a process, however, these ancient approaches to history are abandoned. For Arendt there can be no greatness when history is subsumed to such "laws" as those that govern processes. Here the influence of Nietzsche's monumental history is apparent: "process" and the laws that govern it eliminate uniqueness.[28] This parallels the way that in the sciences the investigation of "processes" has superseded an interest in things according to Arendt ("Concept of History," 57). "Process" as the central figure for understanding history represents the incursion of action-quashing,

---

[26] Walter Benjamin, "The Concept of History," in *Selected Writings*, vol. 4, eds. Howard Eiland and Michael W. Jennings (Cambridge: Harvard University Press, 2006), 389–400.

[27] Arendt is obviously not alone in this characterization of ancient history as a record of greatness, as is clear from Hegel's *Lectures on the Philosophy of World History* and Nietzsche's descriptions of monumental history.

[28] See Judith Sklar, "Rethinking the Past," *Social Research* 44, no. 1 (1977): 80–90.

social-scientific standards into a record of human events – where for Arendt what makes events human and not natural is their occurrence as spontaneous enactments of human freedom, i.e., as genuine "beginnings." Individual acts of greatness – as Homer, Herodotus, and Thucydides once recorded them – become subsumed in our epoch of "process" to laws, a subsumption which deprives those acts of their greatness. Totalitarianism likewise turns human action into precisely such an embodiment of a law of nature or of history (*OT*, 461–465). "Total terror, the essence of the totalitarian government" thereby destroys the human "capacity to make a new beginning" (*OT*, 466). Under totalitarianism, "all action aims at the acceleration of the movement of nature or history" and therefore totalitarianism represents the triumph of "processual" continuity and the eclipse of difference, interruption, and spontaneity (*OT*, 467).

When "process" becomes the central figure of historiography, this represents for Arendt an abasement of freedom. As Frank Kermode points out, Arendt's introduction to *Between Past and Future* explains that "History [...] is a fictive substitute for authority and tradition, a maker of concords between past, present, and future, a provider of significance to mere chronicity."[29] Historiography makes those concords between past, present, and future by smoothing over discontinuity, inventing relevance, and figuring the sequence of events in time as a process. It "historicizes" human action in a sense that is not that of Erich Auerbach – i.e., not in the sense that events are rendered more "realistic" in a terse style, rather it renders that action continuous, non-spontaneous, and heteronomous.[30]

Arendt sees a decline in purely political thinking, beginning in the late eighteenth century, with the representation of human existence as a process and with the new concept of history. Marx is here paradigmatic. He writes about history as a process and, what for Arendt is even worse, as something "made" by human beings, as if history were the end product of a process "of fabrication or making," that comes to a stop when the product is complete ("Concept of History," 79). In accord with the Aristotelian model, in which a process unfolds toward its end, Marx's processual history unfolds toward its utopian end, which devalues and even abolishes what came before. For Arendt Marxist historiography constitutes an index and even an agent of world-alienation, because it values the end over all else and that end *de*values the process which brought it about – that process being the course of all human history. In contrast, storytelling, as the recounting of great deeds and speeches, narrates human freedom,

---

29 Frank Kermode, *The Sense of An Ending: Studies in the Theory of Fiction* (Oxford: Oxford University Press, 2000), 56.
30 Erich Auerbach, *Mimesis: The Representation of Reality in Western Literature* (Princeton: Princeton University Press, 2013), 19.

worldly contingency, and historical possibility without subordinating them to a narrative of process.³¹

The contrast between Arendt's worries about process and Heidegger's epochal worries about historiography as objectifying the past is striking: in Heidegger's 1925 essay on Dilthey, the nineteenth century stands as an ultimate lamentable instance of how academic historiography came to treat the past as an object and thereby occluded the historicity of Dasein.³² For Arendt, historiography in the nineteenth century was lamentably scientific, not in its objectification of the past, but rather in its uptake of the language of development and process that dissolved objects as such ("History," 61). Under the dominance of the figure of process, nothing in modernity is significant by itself and things have significance only as belonging to processes ("Concept of History," 63). Greek and Roman historiography, with their fixation on the lesson of each event, meant that greatness and remembrance for posterity were primary; here again Nietzsche's monumental history is relevant. We in modernity instead model history on natural processes, in which individuality, spontaneity, and freedom are suppressed. What we could call the "totalitarian" element of historiography in modernity is "process" itself insofar as it provides the grounding metaphor of modern historiography. The figural realm of process is part and parcel of the problem of modernity's worldlessness, for Arendt, because it obscures the potential for human action.³³ The Arendtian claim for a unique and extreme form of modern world-alienation, typified in and furthered by the metaphorics of process, provides a counterpoint to Heideggerian fantasies of retrieval. The subordination of historiography to the concept of process threatens the representation of our freedom and thereby the potential of human beings to act freely, do something new, and be a beginning.

---

31 Arendt, *Human Condition*, 181–188. This can be understood as the difference between diegema and diegesis, where the former is an account of episodes and the latter a complete narrative; see Heinrich Lausberg, *Handbook of Literary Rhetoric*, eds. David E. Orton and R. Dean Anderson, trans. Matthew T. Bliss, Annemiek Jansen and David E. Orton (Leiden: Brill, 1998), §1112, 487.

32 Martin Heidegger, "Wilhelm Dilthey's Research and the Current Struggle for a Historical Worldview," in *Becoming Heidegger: On the Trail of His Early Occasional Writings, 1910–1927*, eds. Theodore Kisiel and Thomas Sheehan (Evanston: Northwestern University Press, 2007), 238–274.

33 Hans Blumenberg cites Arendt's claim for the increasing worldlessness of the modern epoch as a counterexample to a more commonplace claim that rise of secularism constitutes the hallmark of modernity. Arendt argues, he points out, that in the modern epoch we are even more alienated from the world than in an epoch of Christian otherworldliness. See Hans Blumenberg, *The Legitimacy of the Modern Age*, trans. Robert M. Wallace (Cambridge: MIT Press, 1983), 8.

Arendt's realist Kantianism

As a work of phenomenology *Origins of Totalitarianism* holds an odd position, for Arendt is not interested primarily in historical origins but in the elements and structures of freedom and totalitarianism. On the other hand, Arendt is attempting to explain totalitarianism as something entirely new and thus historical contextualization is required. The study of totalitarianism takes Arendt into empirical moments where historical events come to pass, but her goal, which concretizes and historicizes Kant's goal in the critiques, is to unfold those moments as actual manifestations of a human free action, where such freedom would be, again in the Kantian sense, a perduring human possibility, albeit one that is threatened by totalitarian worldlessness.[34] Arendt, in other words, wants to inspect the actual, "noumenal" manifestation of human freedom in history. In contrast, Kant's essay on the idea of universal history does not deal in a concrete way with the spontaneity of human freedom, but precisely with an *idea;* he is occupied with the possibility of ideal laws for the aggregate arc of human action.

Arendt maintains a focus on a freedom that would disrupt or interrupt a continuous trajectory of history, the trajectory of process itself, even when that process is defined as governed by a rule of progress. Arendt's interest in a spontaneous freedom that acts without regard to antecedents is, not unlike Benjamin's weak messianic history, much closer to the Kantian discussion of the moral will in the second critique than to Kant's "Idea for a Universal History with a Cosmopolitan Purpose." As we have seen, Kant opens the universal history essay with the question of how we might see human free will on a large scale and writes that in order to do this we need the idea of universal history, where, as we recall, an "idea," which Kant defines as a "concept of reason" is not empirically accessible. Kant claims that we cannot see the free will in history, except if we regard history writ large in accord with the idea of progress.

Arendt, however, wants to observe and promote the *actual* intervention of human freedom and spontaneity in history. At the heart of her Kantian concern for freedom as a spontaneous causality of reason lies a realist infidelity to Kant: her scrutiny of actual historical events in pursuit of a phenomenology of freedom shares no Kantian hesitation over access to the noumena or pure practical reason. She omits the Kantian "as-if" and plainly examines the appearance of freedom in human history. The combination of historical investigation and

---

34 Cf. Hans Vaihinger, *Die Philosophie des Als Ob: System der theoretischen, praktischen und religiösen Fiktionen der Menschheit auf Grund eines idealistischen Positivismus* (Leipzig: Meiner, 1922); Jan Plug, "'As-if' History: Kant, Then and Now" in *Borders of a Lip: Romanticism, Language, History, Politics* (Albany: SUNY Press, 2012).

phenomenological analysis, however, leads to certain contradictions that shape *The Origins of Totalitarianism*: Arendt aims to understand hidden mechanics and subterranean streams, figurations that suggest there are forces at work in this history that are not a matter of freedom. Nonetheless, she expressly strives to write the account of totalitarianism without deducing unprecedented events from their predecessors. She oscillates between insinuating continuity by way of analyzing responsibility, as with anti-Semitism, and making claims for the unprecedented, as with the rise of totalitarianism. The incursion of narrative norms into her historical account illustrates the degree to which narrativity interrupts the phenomenological account that seeks its evidence in the realm of history. For μῦθος, the plot or arrangement of events, constrains even the freedom portrayed in a historical narrative – which means also freedom as such. In the context of Arendt's historical phenomenology, the representation of freedom and its constraints is fraught with these oscillations.

8 Speaking for the Past: On *Begriffsgeschichte* and the Language of Other Epochs

Reinhart Koselleck is less known in U.S. contexts than Heidegger or Arendt, although he belongs to the postwar, and post-Heideggerian, *Nachwuchs* of German historical thought. The span of his work, over five decades beginning in 1954, combines historical, semantic, and phenomenological approaches to the concept of history in order to describe shifts in our relationship to the past and the experience of time. To this end Koselleck investigates the ways that history in Europe has been understood, portrayed, and utilized across epochs. His essays trace alterations in meaning associated with such significant historical concepts as "modernity," "revolution," "crisis," and even "history" itself. Together with Otto Brunner and Werner Conze, Koselleck co-edited the eight-volume lexicon of political and social concepts that is the monument of *Begriffsgeschichte*.[1] Let me note that this German term lends itself to misunderstanding when translated as "conceptual history," which would seem to pertain only to concepts rather than to material reality.[2] Instead, Koselleck's *Begriffsgeschichte* interrogates changes in the meaning of a term in its reciprocal relationship with changing social and political circumstances. *Begriffsgeschichte* deals, in other words, with the historicized meanings of a concept, attending to its sociopolitical contexts and hermeneutic horizons. Koselleck's essays in *Begriffsgeschichte* thus interweave phenomenological, semantic, and historical methodologies into their epochal-hermeneutic narratives.

The individual concepts are not, however, the ultimate focus of interest in Koselleck's oeuvre. Rather, Koselleck aims more broadly, by way of such historico-semantic study, to describe large-scale changes in our connection to the past. His work is thus phenomenological as well as historical and hence it combines the concerns in the foregoing chapters with the oscillation between structural and diachronic analyses. This chapter will argue that his corpus can be seen as developing an account of the modes in which history *speaks for the past*. With its approach to history as a collection of stories, for instance, premodern history spoke for the past in the mode of exemplarity. In modernity, in contrast, "a universal relation of events" was posited, according to which

---

[1] Otto Brunner, Reinhart Koselleck, and Werner Conze (eds.), *Geschichtliche Grundbegriffe: Historisches Lexikon zur politisch-sozialen Sprache in Deutschland* (Stuttgart: Klett-Cotta, 1972–1997).
[2] It also misleadingly evokes *Ideengeschichte* on the model of Ernst Cassirer, or *Geistesgeschichte* on the model of Dilthey.

"history could only speak for itself."[3] In adumbrating these central claims about premodern history and modern historiography, as this chapter will describe in more detail, Koselleck argues that historical writing speaks for the past differently in different epochs. In each case, we will show, the question is one of connection and connectibility: the historiographer connects readers to a time other than their own, according to distinct principles and with respect to different structures of temporality. These tasks of connection and speaking-for proceed by way of historical and semantic investigations of particular concepts in the field of historiography. They also serve as the starting point of a phenomenology of historical time itself. That is, Koselleck comes to describe various formal structures of time in ways that reflect his ties to Kant, the hermeneutic tradition, and Heideggerian analyses of historicity.[4] This includes an inquiry into what he calls the "conditions of possible histories."[5] The result is a transcendental phenomenology of time that is intertwined with the epochal account of a move in historiography from exemplarity to modern historical narrative.

The transcendental-phenomenological strand of Koselleck's thought, we will suggest, is nonetheless not entirely consonant with the periodizing narratives of the *Begriffsgeschichte* for which Koselleck is better known. While Koselleck's phenomenology of time and his epochal *Begriffsgeschichte* of historiography are tightly connected, there are several points of unmarked mutual interference between them. In certain respects, the closing argument of this chapter echoes the previous chapter's investigation of the interference between Arendt's phenomenology of newness and her dependence on historical material. For Koselleck's phenomenological investigation into formal structures of time collides in key respects with his accounts of the different *topoi* by means of which historiography speaks for the past.

Connections to the past

Koselleck's decisive *Begriffsgeschichte* of the concept of history, recounted across numerous essays and articles, traces a terminological modification in German that reflects a modification in the ways that the present connects to

---

[3] Koselleck, "Historia Magistra Vitae: The Dissolution of the Topos into the Perspective of a Modernized Historical Process," in *Futures Past: On the Semantics of Historical Time*, trans. Keith Tribe (Cambridge: MIT Press, 1985), 28.
[4] See Koselleck, "History, Histories, and Formal Structures of Time," in *Futures Past*, 92–104.
[5] Koselleck, *Zeitschichten: Studien zur Historik* (Frankfurt am Main: Suhrkamp, 2000), 99.

and speaks for the past.⁶ The essay on the topos "*historia magistra vitae* [history is the teacher of life]," for example, investigates what sort of teaching history is seen to offer in premodernity, and what sorts of connections it forges between the past and the present. The term *historia*, dating back to Herodotus, had in premodernity designated a report of events. Premodern history was a matter of *historiae*, of collected stories. The stories produced a connection to a past that was approached not as a bygone epoch, but as a repository of lessons or norms. The anecdotes of premodern history spoke for a past that offered ever-valid guidance. Such exemplary history-telling connected the past to the present for the purposes of instruction.⁷

For these reasons exemplary historiography can be understood as a variant of what Roman rhetoric called *synegoria*, referring to the juridical task of advocacy, i.e., of speaking for another. In the juridical scene portrayed by Quintilian, for instance, an orator *speaks for* the *periclitans*, the party who is being judged in a trial, in the operation known as *synegoria*. This is not to say that the orator simply *speaks in the place of* the *periclitans*. Rather, as Rüdiger Campe has argued, the orator crafts the party's message, makes his voice better heard, and in speaking for him, helps the *periclitans* to speak as himself.⁸ Accordingly, we should understand the historiographer as a synegorist, as an advocate of sorts. The historiographer speaks for a past epoch, presenting it effectively in the present, crafting a story for an audience of readers. He submits the past to his audience for judgments of various kinds, with a view toward evoking particular affects and inspiring particular decisions and actions. In this scenario, the past can be seen as occupying a position analogous to that of the ambiguous *periclitans*, for whom and with whom the orator speaks. The historiographer "helps" the past to speak to the present.

In producing a connection to the past and connecting past events to one another in order to render them intelligible, the historiographer depends on such rhetorical devices as hypotyposis – the appeal to the mind's eye – in the vivid evocation of that other time. Such techniques of vivid presentation

---

6 See above all Koselleck, "Geschichte, Historie," in *Geschichtliche Grundbegriffe*, vol. 2 (Stuttgart: Klett-Cotta, 1972), 593–717.
7 For an argument, building on discussions by Hans Blumenberg, that anecdotes are anything but contingent, see Paul Fleming, "On the Edge of Non-Contingency: Anecdotes and the Lifeworld," *Telos* 158 (2012): 21–35; see also Paul Fleming, "'Kannitverstan': The Contingent Understanding of Anecdotes." *Oxford German Studies* 40, no. 1 (2011): 72–81.
8 Rüdiger Campe, "An Outline for a Critical History of Fürsprache: Synegoria and Advocacy," *Deutsche Vierteljahrsschrift für Literaturwissenschaft und Geistesgeschichte* 82, no. 3 (2008): 371.

involve the affectability of both the historiographer and the audience, arousing emotions, or *pathe,* that incline a disposition toward a particular judgment. Campe's goal in studying *Fürsprache,* or advocacy, is ultimately to conceive of a way that we connect to others that would not be bound to a model of primordial individuation.[9] "Affectability" makes a step in this direction, for it denotes a primary openness rather than a primary individuation. Campe points to Aristotle's implications for an originary affectability, or "a world of possible affectedness," that renders us primordially open for connection in the first place by way of the *pathe.*[10] What is important for this chapter is that Koselleck's corpus analyzes the different modes in which history connects to and speaks for the past, thereby evoking different sorts of judgments and *pathe.*

For instance, the premodern conservator of history, in Koselleck's account, crafted the stories from the past for the purpose of persuasion. That conservator drew upon emotions, including fear or admiration, as part of a campaign for a particular judgment by an audience, citing the past in order to evoke a certain behavior or decision in the present. Exemplary anecdotes drawn from the past, in this account, served to offer illustrations of eternal principles relevant to the contemporary audience, rather than simply informing listeners and readers about the past per se. In line with this Cicero, as Koselleck summarizes, treats history as a collection of examples for instruction, Thucydides treats history as a permanent possession that can assist us in acquiring knowledge of similar cases, and Polybius writes that history helps us learn to bear catastrophe by knowing about the catastrophes of others.[11] The fathers of the Roman Catholic church, Koselleck notes, hesitated to allow the anecdotes of profane history to serve as exempla, for this deployment "exceeded the transformatory powers of Church historiography."[12] Nonetheless, Isidore of Seville, Bede, and later Philip Melanchthon made use of "heathen" exemplary stories. In this premodern deployment of history, stories speak for the past as it relates to the present; their exemplary function in effect brooks the difference between past and present.

In the eighteenth-century German-speaking context, the Latin term *historiae* was displaced by *die Geschichte,* understood as the plural form of *das Geschicht* or *das Geschichte.*[13] As evidence of this displacement Koselleck cites the usage of plural verb forms in connection with *die Geschichte* in Baumgarten

---

9 Campe, "Synegoria," 364.
10 Campe, "Synegoria," 373.
11 Koselleck, "Formal Structures of Time," 23, 97.
12 Koselleck, "Historia Magistra Vitae," 24.
13 Koselleck, "Historia Magistra Vitae," 29.

and Herder.[14] Even as the term *historiae* was eclipsed, the premodern sense of history as an additive collection of anecdotes was maintained in the move to *die Geschichte*, taken as a plural noun. In a further development, the plural *die Geschichte* passed into use as a collective singular noun. With that turn to the singular, Koselleck writes, emerged an understanding of history as a unitary entity, rather than as a collection of disparate anecdotes. This shift to a unitary notion of history, which we have observed in Chapter Four's treatment of universal history, defines what Koselleck famously calls the *Sattelzeit*. While the anecdotes of premodern history provided ever-valid lessons, in modernity history was conceived as "a universal relation of events."[15] For Koselleck the very term *die Geschichte*, taken in the singular, combines the two meanings of "a representation of the past" (as in *Historie*) and "the events of the past" (as in the plural of *das Geschicht*). The English word "history" exhibits the same ambiguity, referring both to events and to the representative account that speaks for them. As does the English word "history," the modern singular *die Geschichte* blurs the distinction between these. Hence both the English and German words can refer both to a historiographical account that represents the past in the present and to the events themselves.[16] In this light, Koselleck's *Begriffsgeschichte* can be understood as examining the equivocal semantic zone between historical representation and historical events.[17] The ambiguity of the modern term "die Geschichte" or "history" is therefore not entirely different from the ambiguous position of Cicero's patrician, who speaks both *for* and *as* the defendant, operating between representation and mute presence.

Temporalities and the conditions of history

Heidegger's analyses of historicity form the background of Koselleck's epochal *Begriffsgeschichte*, but how close is Koselleck to Heidegger's epochal history of metaphysics? In a 1985 address on the occasion of Gadamer's birthday, Koselleck

---

14 Koselleck, "Geschichte, Historie," 646.
15 Koselleck, "Historia Magistra Vitae," 28.
16 For White, the eclipse of the representational operations of the historical narrative constitutes the *sine qua non* of narrativity. That is, it can appear as though events simply present themselves in a history, and for this reason, the operations of the author and the aim of the narrative are obscured. See Hayden White, "The Value of Narrativity in the Representation of Reality," in *The Content of the Form: Narrative Discourse and Historical Representation* (Baltimore: The Johns Hopkins University Press, 1987).
17 Campe, "Synegoria," 375.

characterizes his own work as extending Heidegger's analyses of historicity.[18] Koselleck portrays himself as attending to the experience of time throughout history and, drawing more directly on Gadamer, to the "horizons" of expectation and experience. Koselleck pointedly rejects Heidegger's emphasis on "origins, heritage, fidelity, the people, destiny [*Herkunft, Erbe, Treue, Volk, Geschick*]," writing that these concepts "do not adequately ground the conditions of possible histories [*lassen sich die Bedingungen möglicher Geschichten nicht hinreichend begründen*]."[19] Their shortcoming, he claims, is that they do not provide adequate conceptual resources for a consideration of the conditions of possible histories.

It is troubling that Koselleck does not, in that particular context, explicitly repudiate the National Socialist implications of those terms. Indeed, the relationship of his thought to his experience of National Socialism bears further examination.[20] For the purposes of this chapter, what is significant is Koselleck's invocation of what he calls, in numerous essays, the "conditions of possible histories." This phrase indicates that Koselleck's *Begriffsgeschichte* of the concept of history, which emphasizes a move from premodern history to modern history around 1800, also opens onto a transcendental-phenomenological inquiry. This goes along with Koselleck's investigations, in a Heideggerian vein, into what he calls the "formal structures of time" or the "diachronic structures of events."[21] Koselleck's treatments of the formal structures of time can be seen as an elaboration of his epochal *Begriffsgeschichte* of history. For instance, Koselleck argues that the exemplary character of the anecdotes that make up premodern historiography bespeaks a formal structure of time as continuous and an experience of events as repeatable. For this reason there is, in the synegoric procedure of premodern *historia*, no epochal break between past and present, but instead a "relative eternity" that contains all time.[22] In this formal structure of time, the past is continuous with and remains completely relevant

---

[18] Koselleck, "Historik und Hermeneutik," in *Sitzungsberichte der Heidelberger Akademie der Wissenschaften* 1 (1987): 13, translations of this source are mine in each case.
[19] Koselleck, "Historik und Hermeneutik," 13.
[20] Stefan-Ludwig Hoffmann and Sean Franzel, in their introduction to Koselleck, *Sediments of Time: On Possible Histories* (Stanford: Stanford University Press, 2018), ix–xxxi, provide a brief but fascinating discussion of Koselleck's experiences of National Socialism and his views on its commemorations from the 1990s until his death in 2006. Koselleck was conscripted at age seventeen, fought at the eastern front, and was a Soviet prisoner until well after the war had ended.
[21] Koselleck, *The Practice of Conceptual History: Timing History, Spacing Concepts*, trans. Todd Presner, Kerstin Behnke and Jobst Welge (Stanford: Stanford University Press, 2002), 124.
[22] Koselleck, "Historia Magistra Vitae," 25.

to the judgments of the present in offering moral pedagogy. The present need not truly speak *for* the past because the past is fully available.

In contrast, when history is conceptualized in modernity as comprising singular rather than repeatable events, the formal structure of time has changed and past epochs are seen as distinct from the present. As Koselleck writes, with the advent of a linear temporality, a "difference [...] was torn open between one's own time and that of the future, between previous experience and the expectation of what was to come."[23] This tear or breach in time evokes demands within the nascent task of universal history for a principle, ultimately the principle of progress, that could represent the epochs as nonetheless united – even where they seem to present themselves as discrete or disconnected. Time was conceived and experienced differently, namely as internally differentiated and differentiable, as sometimes accelerated and sometimes decelerated.[24] Koselleck's account of the *Sattelzeit* depicts a change in the experience of temporality around 1800 from the cyclical time of a "relative eternity" toward a "genuinely historical temporality."[25] That change in the structure of temporality corresponds to and grounds a move, in Koselleck's account, from a homogeneous temporality of repeatable events toward a discontinuous temporality of singular events.

Koselleck's phenomenological accounts of the formal structures of time ascribe the substantial modifications in how history speaks for the past to a change in the structure and experience of temporality. Prior to the eighteenth century, he explains, time was not historical in a modern sense. This is because history-writing in the premodern epoch inhabited a continuous present. The metaphors history-writing used "impl[ied] a continuous and unbroken present space of experience."[26] There was no need to speak *for* the past because there was no essential estrangement *from* the past in the first place. Each time was the same as each other time and time was not divided into epochs. Only with the advent of modern historiography, so Koselleck's story goes, was the past seen as a different time and were events seen as singular. In this context, the past came to be something historiography needed to "speak for." The new perception of the past as different from the present in turn produced a new demand that historiography create a connection, in order to overcome the gap between past and present. The modern historian came to exist in response to the need for someone to speak for

---

[23] Koselleck, "*Neuzeit:* Remarks on the Semantics of Modern Concepts of Movement," in *Futures Past*, 251.
[24] Koselleck, "Historia Magistra Vitae," 36.
[25] Koselleck, "Historia Magistra Vitae," 30; Koselleck, "Perspective and Temporality: A Contribution to the Historiographical Exposure of the Historical World," in *Futures Past*, 143.
[26] Koselleck, "Perspective and Temporality," 136.

the past, to make that connection. The operation of causal explanation becomes particularly significant with the division of time into past, present, and future, because the historiographer needed to provide a connection between events within a discontinuous temporality.

The thesis that in modernity a shift occurred, away from exemplary history-telling toward epistemological historiography, is not new with Koselleck. What is new is the further phenomenological reflection on that shift in terms of the very structure of temporality.[27] That phenomenological focus on formal structures of time takes on a transcendental dimension when Koselleck formulates an interest in the "conditions of possible histories." It is the subject of some debate among readers of Koselleck what the relationship is between his interest in *Begriffsgeschichte*, formal structures of time, and conditions of possible histories. Stefan-Ludwig Hoffmann understands the conditions of possible histories as the heart of Koselleck's project. He sees it as a study of the phenomenal world rather than historical semantics:

> [C]onceptual history was always only the path and never the goal of his historical thinking. At the center of his thought [...] stood rather the attempt to outline a theory of the conditions of possible histories [...] For Koselleck, linguistic sources always referred to a world beyond the text, to the anthropological, pre-linguistic conditions of historical experience.[28]

In Hoffmann's characterization, Koselleck's *Begriffsgeschichte* always had conditions of possibility as its ultimate concern, rather than any epochal study of history. The *Begriffsgeschichte* would seem here to constitute an entrée to a more

---

27 We could further argue that Koselleck's description of a new sense of history after 1800 both explains and relies upon the birth of epochality itself in European thought. Only with the divisible, nonrepeatable time of modernity is an epoch even thinkable and thus the thinking of epochs defines the epoch in which a break is made with ancient temporality. The introduction of epochality itself into history represents the development of a new "formal structure of time" within Koselleck's history of historiography. According to Koselleck's *Begriffsgeschichte* of *die Geschichte* the emergence of modern historiography reflects the emergence of a new temporality in which epochs were even conceivable to begin with. For premodern temporality, time was everywhere the same and hence there was no epochal division of time at all. The division of time in modern historiography, in contrast, produced an interpretation of time as divisible and nonrepeatable and an interpretation of history as a matter of epochs. There is, we suggest, a recursive logic here that is noteworthy: the interpretation of history as divided into epochs itself defines the modern epoch.

28 Stefan-Ludwig Hoffmann, "Koselleck, Arendt, and the Anthropology of Historical Experience," *History and Theory* 49, no. 2 (2010): 213. Hayden White offers precisely such an assessment of Koselleck as a transcendental or even structuralist philosopher of history in his foreword to *Practices of Conceptual History: Timing, Spacing, Concepts*, by Reinhart Koselleck (Stanford: Stanford University Press, 2002), xii.

significant, transcendental philosophical approach – a legacy of Kant as much as Heidegger. But as Hoffmann points out, the conditions of possible histories are prior to language; they would seem to be a matter of the phenomenal world rather than of textuality. Hoffmann's interpretation suggests that for Koselleck, there was a distinct divide between the textual/linguistic realm and the "world beyond the text." Koselleck's combination of hermeneutic and historical semantics therefore appears, in Hoffmann's reading, not on the side of meaning, but thoroughly on the side of reality.

Helge Jordheim also focuses on the conditions of possible histories in Koselleck, rejecting the characterization of Koselleck as a periodizing thinker but combining his themes of the formal structures of time with the conditions of possible histories.[29] Jordheim argues that Koselleck "seemed eager to distance himself from [periodizing] readings."[30] For Jordheim, Koselleck's project ultimately strives to undermine, rather than to produce, periodization. The epochal narrative is precisely what Koselleck is putting into question. Koselleck executes this questioning, in Jordheim's analysis, in offering a phenomenological account of "times" – not of *time*, but of *times* in the plural, emphasizing the variations and heterogeneities in the formal structures of time and the conditions of possible histories. According to Jordheim, Koselleck's phenomenology of times that inquires into the conditions of possible histories is not only incompatible with the epochalizing narratives of *Begriffsgeschichte*; it even defies the periodizing narrative. Koselleck's goal, in other words, is here understood as subverting periodization rather than producing it.

Narrative and phenomenology: Mutual interferences

Building on Jordheim's analyses, we suggest that in certain respects, Koselleck's *Begriffsgeschichte* collides with his phenomenological, quasi-structural interest in the formal structures of time and the phenomenological-transcendental interest in the conditions of possible histories. Nonetheless, the latter projects do not entirely supplant or supersede the former. Instead, they compete with each other throughout his corpus. Although Koselleck may have aspired to understand the conditions of possible histories according to a Kantian transcendental model, his phenomenological claims for the formal structures of time do not entirely nullify

---

[29] Helge Jordheim, "Against Periodization: Koselleck's Theory of Multiple Temporalities," *History and Theory* 51 (2012): 151.
[30] Jordheim, "Against Periodization," 155.

his epochal account of historiography. Rather, they depend on each other while also undermining each other. The *begriffsgeschichtliche* narrative of the epochs of history is, we suggest, not fully compatible with the quasi-transcendental examination of the formal structures of time, but the phenomenology of the formal structures of times (keeping to Jordheim's formulation) does not exist entirely apart from the periodizing epochal narrative either. While Jordheim sees the phenomenological investigation of "historical times" and conditions of possible histories as Koselleck's genuine goal, the remainder of this chapter will point to mutual interferences between the epochal categories that shape the *Begriffsgeschichte* of history, the phenomenology of time and its formal structures, and the conditions of possible histories, for they evoke elements that they do not fully contain. Specifically, the epochal story of the *Sattelzeit*, with its accounts of the change in how historiography spoke for the past, deconstructs Koselleck's attempts at a phenomenology of time and vice versa, which in turn obscures the conditions of possible histories.

Before examining the incompatibility and mutual interference between the two prongs of Koselleck's enterprise, let us first observe an instance in which the various sides of his project supplement, rather than disrupt, each other. Koselleck's *Begriffsgeschichte* of the concept of chance reflects particularly clearly the epochal turn in historiography in a way that is compatible with its illumination of the significance of the formal structures of time. In "Chance as Motivational Trace in Historical Writing," Koselleck recounts the role played by the concept of chance in premodern historiography and follows its subsequent exclusion from nineteenth-century historiography. In premodern historiography, chance was invoked freely. It appeared in stories of the past in personified form, for instance as the goddess Fortuna.[31] Koselleck argues that the ready reference in premodern historiography to chance reflects the formal structure of time that was the condition of exemplary history. That is, the sameness of time as a "relative eternity" guaranteed "the iterability of all occurrences."[32] The arbitrariness of chance remained a constant, as constant as time itself. Chance fits into this scheme, insofar as fortune's "changeability secured the ever-present preconditions for earthly events and their representability."[33] Hence the exemplary anecdotes of premodern history assumed the actual existence of such caprice and also represented its agency in their tales. Chance and changeability were themselves the constant condition of the sameness of time.

---

31 Koselleck, "Chance as Motivational Trace in Historical Writing," in *Futures Past*, 117–118.
32 Koselleck, "Chance as Motivational Trace," 118.
33 Koselleck, "Chance as Motivational Trace," 118.

In the nineteenth century, in contrast, Koselleck describes how historiographers studiously avoided reference to chance. In this later context, mentions of chance in history unsettled the dominance of the principle of causality. Causes were presumed to connect all events to one another. In modern temporality, any seeming causal discontinuity – which is one way to understand chance – would be perceived as a failure to see, know, or include the cause, rather than as the absence of proximal causality altogether. In a modern history, as Koselleck writes, a description of events as happening merely by chance therefore "indicates an inadequate consistency of given conditions and points to the incommensurability in their results."[34] Here chance stands, in other words, as one name for the phenomenon of events *not* fitting together. It fulfills the narrative demand that a connection be made while defining no particular connection. Chance does not fit into any substantial category of causality; it merely fills a lacuna in cases in which a connection of cause appears to be lacking. The very term "chance," in other words, fills in for an absent connection, for which reason Koselleck calls it a "motivational trace."[35] It marks, but does not smooth over, the discontinuity in a causal sequence of events. In premodern history, on the other hand, the invocation of chance does not threaten the continuity of events, because events are presumed to take place in a homogeneous and ever-same time. In that context, chance appears as an arbitrary arrival of a particular event without vitiating the exemplary function of premodern historiography.

Koselleck's analysis of the concept of chance demonstrates how a *Begriffsgeschichte* can also constitute a phenomenological study of the formal structures of time and the conditions of history-writing. He traces the role and representation of chance across historiographical epochs and elucidates the relevance of that role to the formal structure of time in each epoch. In the context of a premodern temporality that is presumed to be homogeneous, chance can be defined as a genuine occurrence. In the modern context of a temporality that is presumed to be discontinuous, but in which causality everywhere bridges discontinuity, chance names the discontinuity while at the same time providing only an uncomfortably uncertain connection. The *Begriffsgeschichte* of chance and its relationship to historiography illustrates how an epochal account of the concept of chance reflects the formal structure of time that conditions the narration of history. Let us now turn to moments in which Koselleck's attempts at

---

34 Koselleck, "Chance as Motivational Trace," 117.
35 This topic is very close to Campe's work on contingency and probability in *The Game of Probability: Literature and Calculation from Pascal to Kleist*, trans. Ellwood H. Wiggins, Jr. (Stanford: Stanford University Press, 2012).

a phenomenology of time *interfere* with the continuity of Koselleck's narrative of epochal historiography.

Such interference of a phenomenology of times with the epochalizing narrative of history-writing can be seen in the essay "History, Histories, and Formal Structures of Time," in which Koselleck depicts several distinct formal structures of time that belonged to the classical world. In that essay, a variety of temporal structures are said to undergird several distinct arguments for different political constitutions. Koselleck offers the examples of the Persian king Darius the Great (550–486 BCE), Plato's *Laws* (355–347 BCE), and the Greek historian Polybius (200–118 BCE). Darius, as Koselleck shows, represents time as a cogent temporal structure of before and after, earlier and later.[36] This representation of the structure of time suits Darius's account of an evolution toward monarchy in all cases. Plato's *Laws* "sought to derive a periodization of constitutional history from this history itself."[37] Its model of time depends upon on a hypothesis that intervals between phases of rule allowed for learning from previous experience, from which point progress could take place. Polybius, in contrast, bases his writing on a temporal schema of rapid historical decline.[38]

In each of these three premodern writers of history, a general principle of continuity over time is present, but in quite different ways – history is portrayed respectively according to figures of evolution, periodization, and decline. What is significant for the present chapter is that Koselleck's narrative conceit of the *Sattelzeit* and depiction of the premodern temporality interfere with his portrayals of these different formal structures of time in the classical world. First, the analyses of these quite different premodern temporalities unsettles his own argument that there is a single premodern structure of time, one of a "relative eternity" in which events are thoroughly repeatable. Second, their heterogeneity points to the existence in premodernity of nonrepetition and discontinuity, which are supposedly features of historical time *after* 1800. This indicates that exemplarity and iterability cannot be understood univocally to represent premodern history as if in contrast to a more heterogeneous post-1800 historiographical landscape. The epochal narrative of premodern historiography's mode of speaking for the past is incompatible with the different premodern temporalities described in this case.

We see further instances of the mutual interference between epochal history and the phenomenology of time in Koselleck's statements about the

---

[36] Koselleck, "Formal Structures of Time," 96–97.
[37] Koselleck, "Formal Structures of Time," 97.
[38] Koselleck, "Formal Structures of Time," 97.

general nature of history itself. He writes, "History is characterized [...] by the manner in which human foresight, human plans and their execution always diverge in the course of time."[39] Here history is defined phenomenologically, without regard to epochs or narrative conditions, by the divergence of plans and execution, intent and outcome. History is defined, in other words, as waywardness and ungovernability. With this, the very category of history itself is unsettled – for it seems that *nothing* holds it together, that it is instead *defined as* wayward ungovernability. Wayward ungovernability, however, is a quasi-principle even less useful for continuity than that of chance. This deconstructive analysis of history *tout court* as wayward ungovernability contests and is contested by Koselleck's epochal account of premodern and modern history, for the divergence of plans and outcomes that it describes unsettles the distinct formal structures of time that Koselleck has described as characterizing premodern and modern history.

The invocation of temporal discontinuity – as a feature of time per se, rather than of one particular epoch's structure of time – further confounds the distinction between epochal history and the phenomenology of time. The philosophical claim just cited for ubiquitous temporal discontinuity and waywardness dissolves the distinction between epochs. For Koselleck implies, without distinguishing premodern and modern history, that *all* historiography covers over fundamental and profound disconnections in time:

> Jede rückläufige Deutung zehrt von einem Geschehen in seinem vergangensein, das in jeweiligem Heute neu zur Sprache gebracht wird. Eine Geschichte geht also in den vielschichtig gebrochene Zeitverlauf ein, in dem sie, bewußt oder unbewußt tradiert, immer wieder neu artikuliert wird. (Koselleck, "Terror und Traum: Methodologische Anmerkungen zur Zeiterfahrungen im Dritten Reich," *Vergangene Zukunft*, 282)

> Each retrospective interpretation feeds off the pastness of an occurrence and seeks to articulate it anew in the present. A history thus enters a complexly fractured temporal succession and is continually rearticulated, whether consciously or unconsciously handed down. (Koselleck, "Terror and Dream: Methodological Remarks on the Experience of Time during the Third Reich," in *Futures Past*, 216)

Here Koselleck describes temporal succession as "fractured" and yet, the fracture is patched over by a rearticulation. Historiography as constant rearticulation is portrayed as a renewing and reconnecting reception through and through. That renewing reception, however, depends upon a temporal disconnection produced by the fractures of temporal succession. Times, in other

---

[39] Koselleck, "On the Disposability of History," in *Futures Past*, 207.

words, would seem to be heterogeneous with respect to each other without respect to epoch, to be divided by no definitive and profound breaks, and instead to be connected to each other as rearticulations.

On the other hand, Koselleck's phenomenological representation of time as a matter of *Zeitschichten*, or layers of time, complicates his suggestions that temporal succession is thoroughly fractured. The figurality of *Zeitschichten* would seem to mitigate the otherness of times to one another while undermining any epochal narrative. Specifically, Koselleck claims that *Zeitschichten*, or strata of time, operate at different rates of change.[40] The implication is that there would be no *total* otherness of other times, insofar as the complex layering of time itself connects, albeit indirectly, our times to others and in principle all times to other times in variously mediated fashions. That layering of time means that some moments are connected in some ways and all are *somehow* connected, even if indirectly, to others. This partial and variegated connection suggests that there is no single governing principle of continuity or discontinuity. The model of *Zeitschichten* also suggests that epochs may be contiguous and yet without connection. This sheer juxtaposition clearly constitutes a deconstructive counterpoint to the epochal narrative of history. Furthermore, the thesis of *Zeitschichten* provides a phenomenology of time that, once again, is transepochal. It therefore speaks against Koselleck's conceptual history of history and its claims for a shift in the *experience* of temporality with an eye instead to the *real* conditions of history.

The discontinuity of *Zeitschichten* is even more pronounced when Koselleck, using a phrase that originated with Ernst Bloch, defines the epoch-confounding phenomenon of the "contemporaneity of the noncontemporaneous [*Gleichzeitigkeit der Ungleichzeitigen*]."[41] In explaining this, Koselleck offers the concrete example that an epochal step in technology may unfold in one place while in another place there is only minor technological advancement.[42] This contemporaneity of the noncontemporaneous constitutes a strike against universal history, because it represents a dislocation and nonhomogeneity within a present moment itself.[43]

---

40 Koselleck, *Zeitschichten*, 238. See also John Zammito, "Koselleck's Philosophy of Historical Time(s) and the Practice of History," in *History and Theory* 43 (2004): 130.
41 Ernst Bloch, "Nonsynchronism and the Obligation to its Dialectics," trans. Mark Ritter, *New German Critique*, no. 11 (1977): 22–38.
42 Koselleck, "The Eighteenth Century as the Beginning of Modernity," in *The Practice of Conceptual History*, 159.
43 Of course this phenomenological insight into time does not originate with Koselleck. Herder, for one, is significant in disavowing the singularity of time and considers whether

The contemporaneity of the noncontemporaneous has implications that extend beyond the notion that epochs in history are not homogeneous and unimpeachably identifiable. It means, more radically, that what we are doing at any moment is also not homogeneous and unimpeachably identifiable. We are all, in other words, always doing more than we know we are doing. Different things, moments, and trajectories exist together, exceeding identitarian fantasies of epochality. The contemporaneity of the noncontemporaneous is thus a source of deep undecidability. It is ultimately not even temporal in any conventional sense, insofar as it refers to latency and possibility. It constitutes a condition for *any* possible history, rather than for particular histor*ies*, for it means that in every moment elements abide that do not belong to that moment. Put otherwise, the contemporaneity of the noncontemporaneous suggests that the definition of a particular moment requires the repression of its heterogeneous others. It in principle names the overdetermination of and in every moment. The thesis of the contemporaneity of the noncontemporaneous works against Koselleck's claims for the move from exemplarity to epochality, insofar as that claim defines a "before" and an "after" in an epochal mode. The deconstructive thrust of the thesis disrupts Koselleck's epochal account of premodernity, *Sattelzeit*, and modernity. On the other hand, the phenomenological endeavor also *depends* on that epochal history whose principles of organization it undermines.

Koselleck's exemplary phenomenology

Even within Koselleck's epochal narrative of history there are mutual interferences of elements. Koselleck's claims for the significance both of exemplarity in premodernity and of the unrepeatability of events in modernity reflect an inadvertent tension in the epochal periodization of historiography. Koselleck's periodization, as we have seen, characterizes the premodern formal structures of time, in which history had served as a reservoir of examples within an ever-same time, as making way for a new temporality around 1800. The loss of the ordering principle of exemplarity made way for universal history's claims for large-scale development over time and for structures of time that operate beyond the scope of single events in history. This new time of modernity [*Neuzeit*] comes to be ordered by the concept of progress, as in Kant's "Idea for a Universal History with a Cosmopolitan Purpose." According to Koselleck, the

---

there are as many times as objects. See Zammito, "Koselleck's Philosophy of Historical Time (s)," 125. Leibniz and calculus are also recognizable in this gesture toward the infinitesimal.

historical trope of progress estranged the past and privileged the unknown future.[44] The concept of progress, precisely as Kant demonstrates, served to unify the historical data and make sense of a non-exemplary, nonrecursive time. There cannot be, it seems, narrative history without a concept that guides it, whether that concept is a past-oriented exemplarity or a future-oriented progress. This means that the concept of progress gives history a coherence and even wholeness when exemplarity as a principle seems to dissolve.

We should note, however, that the concept of progress does not void the significance of exemplarity. On the contrary, the concept of progress demands that single events be taken as *exemplary* of large-scale developments. Despite Koselleck's claims that the principle of exemplarity belongs to premodernity, modern historiography's reliance on a concept, and in particular the concept of progress, renders modern historiography *also* a form of exemplarity. For the unitary notion of a progressive history de facto recasts individual events as *examples* of progress. On this point Koselleck quotes Frederick the Great at the cusp of the turn from exemplary to progressive historiography, to the effect that with history, the incidents remain the same, only names change.[45] This demonstrates that the iterable character of premodern history abides in the progress that structures a narrative of universal history. Koselleck also quotes Frederick's declaration that no one learns from examples drawn from the past; this appears to be a consummately modern rejection of exemplary history.[46] Nonetheless, Koselleck points out that the very thesis "no one has learned from examples from the past" would seem itself to be a lesson drawn from instances in which examples were not learned from. Any instances of not-learning from an example would thereby serve as examples of Frederick's thesis that no one learns from examples![47] Likewise even the notion of epochality harbors a principle of exemplarity, for it implicitly or explicitly casts individual events as exemplary of the epoch to which they are said to belong.[48]

---

44 Koselleck, "Chance as Motivational Trace," 127.
45 Koselleck, "Historia Magistra Vitae," 26.
46 Koselleck, "Historia Magistra Vitae," 26.
47 Koselleck, "Historia Magistra Vitae," 26. Likewise White's study of modes of emplotment in *Metahistory* declares that exemplarity remains in force even within epochal temporality, insofar as nineteenth-century European historiographical accounts can be classified as examples of a particular genre.
48 Hayden White's work on narrativity makes several other significant points for the question of exemplarity and speaking-for in historiography. White distinguishes between narration, which acknowledges the subject-position of the narrator, and narrativization, which is that form of discourse that feigns objectivity and disavows representation, as if there were no one speaking, as if the events told themselves ("Value of Narrativity," 3). Narrativized historiography

This concluding consideration of exemplarity fits with Koselleck's *Begriffsgeschichte* in which premodern time knew no singular events *and* it also illustrates the conundrums of delimiting the formal structures of time and the conditions of possible histories. It shows that the concepts of the contemporaneity of the noncontemporaneous, and of *Zeitschichten*, do not singlehandedly undermine an epochal determination of Koselleck's *Begriffsgeschichte*. Rather, the *Begriffsgeschichte* is already imbued with proto-deconstructive elements that destabilize its seeming univocality. What is more, both *Zeitschichten* and the contemporaneity of the noncontemporaneous depend upon the epochal presuppositions they seem to undermine. Within the concrete analyses expressed in his conceptual histories, Koselleck develops a repertoire of quasi-transcendental concepts that undermine epochality. His methodological proposals ultimately unsettle both his own claims for epochality and, in a broader sense, the notion of epochality as such. This may be seen as his deeper debt to, if not Heidegger, then to a deconstructive post-Heideggerian legacy. For the quasi-structural analyses of historical temporality itself are not fully consonant with the epochal story that Koselleck tells of the changes in the function and form of historical writing.

---

disavows, we could say, the operation of historiographical synegoria, i.e. of history as speaking for another time. It pretends that there is no one speaking, that there is an objective and immediate self-speaking or self-presentation of the events without any synegoric interference. This supposedly non-synegoric narrativity depends upon multiple formal concessions and produces a variety of effects. The agenda, techniques, and even writerly agency that gave rise to the narrative are obscured by the seeming naturalness of the self-presentation of the events recounted. Precisely in that obscurity, however, narrativizing always harbors a form of moralizing, according to White. Above all, while the events may seem to speak for themselves, the closure that ensures the completeness of the narrativized account of events harbors a moral, defining and sealing the significance of the recounted events. For White, narrativized historiography always therefore is "speaking for" a moral. This means, in turn, that events can only be narrated historically at all when they are seen as exemplary, i.e., as speaking-for a larger principle. This is true even within the modern epoch, in which exemplarity has supposedly fallen away as a governing temporo-historical principle.

Conclusion: Wholeness and its Sabotage

As we have seen in the foregoing chapters, we formulate and comprehend narratives with an expectation of unity and wholeness. According to Koselleck, modern historiography emerged in an attempt to see history as a unitary whole in the context of a change in the formal structure of time; when the concept of exemplarity dissolves around 1800, a new concept fills in, such as progress, decline, or universality. The unity of a whole is a quintessential issue for the development of aesthetics. Hence in the wake of Baumgarten, and at the point of the emergence of modern historiography, Kant claims that the cognition of a whole as a unity is performed by the imagination, along with the operation of understanding, in assigning a concept to the manifold of perception. History newly conceived as a whole, as universal history in the singular, requires an abstraction under which to subsume particular events and even the differentiable layers of time itself. The knowledge of history as a whole, in other words, takes its clue from nascent aesthetic theory and its investigation of how cognition requires a synthesis and produces a whole.

Aristotle's prescriptions for plot in the *Poetics* are likewise focused on the production of a complete and integral plot, in which the events are properly connected. The incompatible and contradictory interpretations of the *Poetics*, across the history of literary theory, testify to the potentiality it holds for considerations of necessity, contingency, and connection. In *Reading for the Plot*, Peter Brooks argues that plot is temporal and meaning-oriented, describing plots as "temporal syllogisms, concerning the connective processes of time."[1] Here connection over time is crucial: the "logic" of plot is intertwined with temporality, with "those messages that are developed through temporal succession."[2] Plot, precisely as a logic, involves a "principle of interconnectedness and intention which we cannot do without in moving through the discrete elements – incidents, episodes, actions – of a narrative."[3] Brooks thus emphasizes the temporal, successive, and dynamic aspects of plot.[4] The argument for a temporalized plot rejects the

---

[1] Peter Brooks, *Reading for the Plot: Design and Intention in Narrative* (New York: Vintage Books, 1984), 21.
[2] Brooks, *Reading for the Plot*, 10,
[3] Brooks, *Reading for the Plot*, xi, 5. Likewise antinarrative does not throw a wrench into Brooks' analysis, for even "in the absence of pattern and structure, patterning and structuration remain necessary projects, dynamic intentions" (311). With modernist and postmodernist narrative, therefore, the reader is "forced to engage in plotting, if not toward the creation of meaning, at least in exploration of the conditions of narrative meaning" (316).
[4] Brooks, *Reading for the Plot*, 18–19.

relative stasis of traditional structuralist narratology. Plot is instead defined as a "structuring *operation*," a "logic and *dynamic* of narrative," and an "*active* shap*ing* force."[5] The reconception of plot as a dynamic and active operation evokes a temporal principle of connection. For this reason Brooks favors the polysemy of the English word "plot" over Aristotle's μῦθος, the latter of which seems, certainly in its reception, to refer to a structure made up of static elements. If we combine Blumenberg's invocations of *Kontext* in "The Concept of Reality and the Possibility of the Novel" with Brooks's sense of a dynamic plot, we might say that with regard to narrative, reality's fitting-together unfolds temporally and actively as the plot itself.

As we have seen, however, apart from the unity that plot entails in defining an arc or trajectory over time with a defined beginning and ending, there is also a principle of connection that operates *between* elements one by one. These "interstitial" connections are also necessary in defining a larger whole, for they relate the elements to one another and give the clue to – or even the substance of – the meaningfulness of what is being narrated. To state the matter another way, form in the broadest sense is a matter of unity and completeness, which de facto controverts the representation of discontinuity, randomness, or mere one-by-one connection of events. On the other hand, insofar as narratives evoke human freedom and worldly contingency, the continuity of events cannot be portrayed as entirely necessary. To an extent, therefore, the representation of contingent, one-by-one connections between individual events potentially interferes with or subverts the formal unity of a narrative. There is, as we have seen, an insuperable tension between the formal unity of narrative and the representation of spontaneous, free, contingent, or underdetermined action within narratives. This tension is seen in Gottsched's attempts to explain the connections of poetic thinking, in Kant's redefinition of *cognitio historica*, in Arendt's narrative of totalitarianism, and in Koselleck's historicization of the structures of temporality. Even the very notion of historicity [*Geschichtlichkeit*] in Heidegger and the goal of expressing "how it really was" in Ranke pose dilemmas to definition and narration, for epochal claims both connect moments and announce particular disconnections or discontinuities between spans of time. In contrast, the inexorable

---

[5] Brooks, *Reading for the Plot*, 10 (emphasis mine); 13. Formal analysis and narratology, for Brooks, are in this respect not enough focused on the dynamic shaping force of plot; Gérard Genette's pseudo-time, for instance, which makes use of the time of reading to present a span of time, is helpful but ultimately not dynamic enough to explain the shaping force of plot in Brooks' broader sense (20). For Brooks the sense of "plot" that pertains to conspiracy rightly evokes the intentional, forward-moving aspect of literary plot (12).

connections of the tragic events in Aristotle and in Kant's "novelistic" universal history each require a structural unity that holds together the separate elements from "outside," as it were.

In each case that we have observed in the preceding chapters, narrative connections are proposed that cannot be guaranteed, although they are demanded by the particular text in question. Hence, we have examined the conflict between overarching coherence and one-to-one connection in prescriptions for poetic narrative, in representations of human freedom in nascent historiography, and in epochal claims in phenomenology. For the plotting of history – the way a unitary history is constituted with individual connections between events – is by definition a narrative procedure. Conversely, every literary narrative contains a historical element, at least insofar as it involves an account of a second-order actuality within a possible world.[6]

Apart from the cognitive requirement for wholeness in achieving narrative intelligibility, Hayden White argues that there exists another impetus toward unitary narrative. Specifically, he asks what wish is enacted in the narrativity thesis, i.e., in the fantasy that events, on their own, exhibit a formal coherence.[7] He proposes that narrativity and the closure it requires speak to our desires, indeed that the completeness and wholeness that narrativity offers is itself the object of our desire. (We observe here a surprising echo of Meier's Wolffian claim, as discussed in Chapter Three, that a perfect historical knowledge is beautiful.) The closure of an ending, for instance, provides a sense of wholeness that we ourselves lack.[8] Because of this, "the reality represented in the historical narrative, in 'speaking itself', speaks to us, [...] and displays to us

---

[6] The constellation of possible worlds theories, the development of the novel, and the emergence of modern historiography is well known. The emergence of modern historiography as a discipline in Germany has been said to happen only thanks to the infusion of novelistic aesthetics into history-writing, see Chapter Four and, for instance, Frank R. Ankersmit, "The Linguistic Turn, Literary Theory and Historical Theory," *Historia* 45, no. 2 (2000): 271–310; Daniel Fulda, *Wissenschaft aus Kunst: Die Entstehung der modernen deutschen Geschichtsschreibung 1760–1860* (Berlin, De Gruyter, 1996); and Jörn Rüsen, "Rhetoric and Aesthetics of History: Leopold von Ranke," *History and Theory* 29, no. 2 (1990): 190–204. See also Catherine Gallagher, "The Rise of Fictionality," in *The Novel*, vol. 1, ed. Franco Moretti (Princeton: Princeton University Press, 2006), 336–363; and Gallagher, "War, Counterfactual History, and Alternate-History Novels," *Field Day Review* 3 (2007): 52–65.
[7] Hayden White, "The Value of Narrativity in the Representation of Reality," in *The Content of the Form: Narrative Discourse and Historical Representation* (Baltimore: The Johns Hopkins University Press, 1987), 4.
[8] White, "Value of Narrativity," 24.

a formal coherency to which we ourselves aspire."[9] We desire and aspire to a fantasmatic formal coherence, a completion that, as Heidegger's Dasein reflects, we can never achieve and that Arendt would define as the foreclosure of new beginnings and thus of freedom. History would seem to make the real desirable, however, in portraying the events of the past as complete and coherent. It also produces a conflict between the fantasmatic wholeness that history displays and that we desire and the real, which does not possess within itself such formal coherence.

White's description of the desire for wholeness brings us back to the *pathe* that were mentioned in the context of Koselleck's synegoric historiography, namely in terms of a "fundamental affectibility" that we share. What, we might ask, is the *pathos* of desire for wholeness? Or is desire for wholeness the condition for *all* the *pathe*, is it the condition of affectibility in the first place? If the latter is the case, i.e., if desire for wholeness is the condition for the *pathe* that constitute affectibility in the first place, then that desire for wholeness would not be one *pathos* among others but the origin of them all. The desire for wholeness would constitute the absolute condition of the "world of possible affectedness."[10] This would explain why the Aristotelian passions, as Campe points out, are blueprints for different kinds of narratives.[11] They ground such different kinds of narratives because they are different expressions in different *pathe* of the singular founding *pathos* of desire for wholeness. They represent possibilities – affectabilities – which themselves derive from a single basic affectibility that is the desire for wholeness. This would mean that as a proxy for the meaningfulness that grants wholeness, narrative form is itself an object of desire. Narrative form gives coherence by making a whole out of disconnected events and yet, it also constitutes the sum of their *dis*connections.

Let us conclude with a question based on this assumption that we desire the wholeness that we bring to historical narrative, literary narrative, and narrative in general – that we desire in that respect form itself. If it is the case that we desire in narratives the wholeness that we ourselves lack, then how are we to understand the desire to *sabotage* the appearance of wholeness, or at least to unmask its ideological, fantasmatic character? This is where the inquiry becomes reflexive: this entire inquiry has, in effect, unmasked the fabulated, even fantasmatic character of a diverse set of forms of connection that make for

---

9 White, "Value of Narrativity," 21.
10 Rüdiger Campe, "An Outline for a Critical History of Fürsprache: Synegoria and Advocacy." *Deutsche Vierteljahrsschrift für Literaturwissenschaft und Geistesgeschichte* 82, no. 3 (2008): 369.
11 Campe, "Synegoria," 370.

particular unitary narratives in the authors we have surveyed. The argument for a mutual interference between overarching plot and one-by-one connection amounts to precisely such a sabotage. In the artful possibilities for producing historical completeness, we see the eclipse of one-by-one contingency and disconnection. In the wayward ungovernability of poetic thoughts or even history itself, we see the defeat of a fantasmatic wholeness. While a particular narrative may portray itself as complete, however, others emerge in response to its incontrovertible incompleteness, which derives in turn from the incompleteness of the desire that gave rise to it, and from the never-finished dynamic unfolding of the conditions of temporality that render perfect coincidence and completeness a fiction.

Bibliography

Adorno, Theodor. *History and Freedom: Lectures 1964–1965*. Edited by Rolf Tiedemann. Translated by Rodney Livingstone. Cambridge: Polity Press, 2006.
Adorno, Theodor. *Zur Lehre von der Geschichte und der Freiheit 1964–1965*. Frankfurt am Main: Suhrkamp, 2001.
Allen, Wayne. "Hannah Arendt and the Ideological Structure of Totalitarianism." *Man and World: An International Philosophical Review* 26, no. 2 (1993): 115–129. https://doi.org/10.1007/BF01273194.
Allison, David. "Spontaneity and Autonomy in Kant's Conception of the Self." In *The Modern Subject: Conceptions of the Self in Classical German Philosophy*, edited by Karl Ameriks and Dieter Sturma, 11–30. Albany: SUNY Press, 1995.
Ameriks, Karl. *Kant and the Historical Turn: Philosophy as Critical Interpretation*. Oxford: Oxford University Press, 2006.
Anderson, John G.T. *Deep Things Out of Darkness: A History of Natural History*. Berkeley: University of California Press, 2013.
Ankersmit, Frank R. "The Linguistic Turn, Literary Theory and Historical Theory." *Historia* 45, no. 2 (2000): 271–310.
Ankersmit, Frank R. *Sublime Historical Experience*. Stanford: Stanford University Press, 2005.
Ankersmit, Frank R. "White's 'Neo-Kantianism': Aesthetics, Ethics, and Politics." In *Refiguring Hayden White*, edited by Frank R. Ankersmit, Ewa Domanska, and Hans Kellner, 34–53. Stanford: Stanford University Press, 2009.
Arendt, Hannah. "The Concept of History: Ancient and Modern." In *Between Past and Future: Eight Exercises in Political Thought*, 41–90. New York: Penguin, 1968.
Arendt, Hannah. "Europe and the Atomic Bomb." In *Essays in Understanding 1930–1954: Formation, Exile and Totalitarianism*, edited by Jerome Kohn, 418–422. New York: Harcourt, Brace & Co., 1994.
Arendt, Hannah. *The Human Condition*. Chicago: University of Chicago Press, 1998.
Arendt, Hannah. *On Revolution*. New York: Penguin, 1963.
Arendt, Hannah. *Origins of Totalitarianism*. New York: Harcourt, 1976.
Arendt, Hannah. *Rahel Varnhagen: The Life of a Jewess*. Baltimore: The Johns Hopkins University Press, 1997.
Arendt, Hannah. "A Reply to Eric Voegelin." In *The Portable Hannah Arendt*, edited by Peter Baehr, 157–164. New York: Penguin, 2000.
Aristotle. *Art of Rhetoric*. Translated by J.H. Freese. Cambridge: Harvard University Press, 2006.
Aristotle, *Metaphysics*. Translated by Hugh Tredennick. Cambridge: Harvard University Press, 1933.
Aristotle. *Physics*. Translated by P.H. Wickstead and F.M. Cornford. Cambridge: Harvard University Press, 1980.
Aristotle. *Poetics*. Translated by Leon Golden. Tallahassee: Florida State University Press, 1981.
Aristotle. *Poetics*. Translated by Stephen Halliwell. Cambridge: Harvard University Press, 1995.
Aristotle. *Poetics*. Translated by Richard Janko. Indianapolis: Hackett Publishing Co., 1987.
Aristotle, *Poetics*. Translated by W. Rhys Roberts. London: Heinemann, 1927. http://data.perseus.org/citations/urn:cts:greekLit:tlg0086.tlg034.perseus-eng.
Armstrong, J.M. "Aristotle on the Philosophical Nature of Poetry." *The Classical Quarterly* 48 (1998): 447–455. https://doi.org/10.1093/cq/48.2.447.

Auerbach, Erich. *Mimesis: The Representation of Reality in Western Literature*. Princeton: Princeton University Press, 2013.
Baehr, Peter. Editor's introduction to *The Portable Hannah Arendt*, vii–liv. New York: Penguin, 2000.
Bahti, Timothy. *Allegories of History: Literary Historiography after Hegel*. Baltimore: The Johns Hopkins University Press, 1992.
Bambach, Charles R. *Heidegger, Dilthey and the Crisis of Historicism*. Ithaca: Cornell University Press, 1995.
Bambach, Charles R. *Heidegger's Roots: Nietzsche, National Socialism, and the Greeks*. Ithaca: Cornell University Press, 2003.
Barash, Jeffrey Andrew. *Martin Heidegger and the Problem of Historical Meaning*. New York: Fordham University Press, 2003.
Bartky, Sandra Lee. "Heidegger and the Modes of World-Disclosure." *Philosophy and Phenomenological Research* 40, no. 2 (1979): 212–236. https://doi.org/10.2307/2106318.
Bartuschat, Wolfgang. "Kant über Grundsatz und Grundsätze in der Moral." *Jahrbuch für Recht und Ethik* 12 (2004): 283–98. http://www.jstor.org/stable/43593198.
Battini, Michele. "The Birth of an Anti-Jewish Anti-Capitalism." *Constellations: An International Journal of Critical and Democratic Theory* 16, no. 4 (2009): 615–633. https://doi.org/10.1111/j.1 467-8675.2009.00567.
Beiser, Frederick C. *Diotima's Children: German Aesthetic Rationalism from Leibniz to Lessing*. Oxford: Oxford University Press, 2009.
Beiser, Frederick C. *The German Historicist Tradition*. Oxford: Oxford University Press, 2011.
Beiser, Frederick C. "Hegel and Ranke: A Re-Examination." In *A Companion to Hegel*, edited by Stephen Houlgate and Michael Baur, 332–351. Oxford: Wiley-Blackwell, 2011.
Beiser, Frederick C. "Hegel's Historicism." In *The Cambridge Companion to Hegel*, edited by Frederick C. Beiser, 270–300. Cambridge: Cambridge University Press, 1993.
Bell, Matthew. *The German Tradition of Psychology in Literature and Thought, 1700–1840*. Cambridge: Cambridge University Press, 2005.
Benjamin, Walter. "The Author as Producer." In *Selected Writings*, volume 2, no. 2, edited by Michael W. Jennings, Howard Eiland, and Gary Smith, 768–782. Cambridge: Harvard University Press, 2005.
Benjamin, Walter. "Der Autor als Produzent." In *Gesammelte Schriften*, vol. 2, no. 2, 683–701.
Benjamin, Walter. "The Concept of History." In *Selected Writings*, vol. 4, edited by Howard Eiland and Michael W. Jennings, 389–400. Cambridge: Harvard University Press, 2006.
Benjamin, Walter. *The Correspondence of Walter Benjamin, 1910–1940*. Edited by Gershom Scholem and Theodor Adorno. Translated by Manfred R. Jacobson and Evelyn M. Jacobson. Chicago: University of Chicago Press, 1994.
Benjamin, Walter. *Gesammelte Briefe*, vol. 1. Edited by Theodor Adorno and Gershom Scholem. Frankfurt am Main: Suhrkamp, 1978.
Benjamin, Walter. *Gesammelte Schriften*. Edited by Rolf Tiedemann and Hermann Schweppenhäuser, 7 vols. Frankfurt am Main: Suhrkamp, 1972–1989.
Benjamin, Walter. "Review of Hönigswald's *Philosophie und Sprache*." In *Selected Writings*, vol. 4., edited by Howard Eiland and Michael W. Jennings, 139–144. Cambridge: Harvard University Press, 2006.
Benjamin, Walter, "Richard Hönigswald, *Philosophie und Sprache: Problemkritik und System*." In *Gesammelte Schriften*, vol. 3, 564–569.

Benjamin, Walter. "Über den Begriff der Geschichte." In *Gesammelte Schriften*, vol. 1, no. 2, 690–708.
Benjamin, Walter and Theodor Adorno, *The Complete Correspondence, 1928–1940*. Cambridge: Polity, 1999.
Bennett, Benjamin. *Beyond Theory: Eighteenth-Century German Literature and the Poetics of Irony*. Ithaca: Cornell University Press, 1993.
Berghahn, Klaus L. "German Literary Theory from Gottsched to Goethe." In *The Eighteenth Century*, vol. 4 of *Cambridge History of Literary Criticism*, edited by H.B. Nisbet and Claude Rawson, 522–545. Cambridge: Cambridge University Press, 2005.
Bernstein, Richard J. *Hannah Arendt and the Jewish Question*. Cambridge: MIT Press, 1996.
Bernstein, Richard J. *The New Constellation: The Ethical-Political Horizons of Modernity/Postmodernity*. Cambridge: Polity Press, 1991.
Birke, Joachim. "Gottscheds Neuorientierung der deutschen Poetik und der Philosophie Wolffs." *Zeitschrift für deutsche Philologie* 85 (1966): 560–575.
Birmingham, Peg. *Hannah Arendt and Human Rights: The Predicament of Common Responsibility*. Bloomington: Indiana University Press, 2006.
Blackall, Eric. *The Emergence of German as a Literary Language, 1700–1775*. Ithaca: Cornell University Press, 1978.
Blattner, William D. "Heidegger and Philosophical Modernism." *Inquiry: An Interdisciplinary Journal of Philosophy* 38, no. 3 (1995): 257–276. https://doi.org/10.1080/00201749508602389.
Bliss, Mavis Louise. "Arendt and the Theological Significance of Natality." *Philosophy Compass* 7, no. 11 (2012): 762–771. https://doi.org/10.1111/j.1747-9991.2012.00515.x.
Bloch, Ernst. *Erbschaft dieser Zeit*. Frankfurt am Main: Suhrkamp, 1973.
Bloch, Ernst. "Nonsynchronism and the Obligation to its Dialectics." Translated by Mark Ritter. *New German Critique*, no. 11 (1977): 22–38. doi:10.2307/487802.
Bloom, Harold. *The Anxiety of Influence: A Theory of Poetry*. Oxford: Oxford University Press, 1997.
Blumenberg, Hans. "The Concept of Reality and the Possibility of the Novel." Translated by David Henry Wilson. In *New Perspectives in German Literary Criticism: A Collection of Essays*, edited by Richard E. Amacher and Victor Lange, 29–48. Princeton: Princeton University Press, 1979.
Blumenberg, Hans. "Lebenswelt und Technisierung unter Aspekten der Phänomenologie." *Filosofia* 14 (1963): 855–884.
Blumenberg, Hans. *The Legitimacy of the Modern Age*. Translated by Robert M. Wallace. Cambridge: MIT Press, 1983.
Blumenberg, Hans. *Die Legitimität der Neuzeit*. Frankfurt am Main: Suhrkamp, 1966.
Blumenberg, Hans. *Paradigmen zu einer Metaphorologie*. Frankfurt am Main: Suhrkamp, 1999.
Blumenberg, Hans. *Paradigms for a Metaphorology*. Translated by Robert Savage. Ithaca: Cornell University Press, 2010.
Blumenberg, Hans. "Wirklichkeitsbegriff und Möglichkeit des Romans." In *Nachahmung und Illusion: Kolloquium Gießen, Juni 1963*, edited by Hans Robert Jauss, 9–27. Munich: Fink, 1964.
Boeder, Heribert. "Heidegger's Legacy: On the Distinction of 'Aletheia.'" Translated by Marcus Brainard. *Research in Phenomenology* 28 (1998): 195–210.
Boym, Svetlana. *Another Freedom: The Alternative History of an Idea*. Chicago: University of Chicago Press, 2010.

Brobjer, Thomas. "Nietzsche's Relation to Historical Methods and Nineteenth-Century German Historiography." *History and Theory* 46, no. 2 (2007): 155–179. https://doi.org/10.2307/4502236.

Brodsky, Claudia. "Lessing and the Drama of the Theory of Tragedy." *MLN* 98, no. 3 (1983): 426–453. https://doi.org/10.2307/2906018.

Brooks, Peter. *Reading for the Plot: Design and Intention in Narrative*. New York: Vintage Books, 1984.

Brunkhorst, Hauke. "The Origins of Totalitarianism/Elemente und Ursprünge totaler Herrschaft." In *Arendt-Handbuch: Leben, Werk, Wirkung*, edited by Wolfgang Heuer, Bernd Heiter and Stefanie Rosenmüller, 35–42. Stuttgart: J.B. Metzler, 2011.

Brunner, Otto, Reinhart Koselleck, and Werner Conze (eds.). *Geschichtliche Grundbegriffe: Historisches Lexikon zur politisch-sozialen Sprache in Deutschland*. Stuttgart: Klett-Cotta, 1972–1997.

Burke, Peter. "Ranke the Reactionary." In *Leopold von Ranke and the Shaping of the Historical Discipline*, edited by Georg G. Iggers and James M. Powell, 36–44. Syracuse: Syracuse University Press, 1990.

Campe, Rüdiger. *Affekt und Ausdruck: Zur Umwandlung der literarischen Rede im 17. und 18. Jahrhundert*. Tübingen: Niemeyer, 1990.

Campe, Rüdiger. *The Game of Probability: Literature and Calculation from Pascal to Kleist*. Translated by Ellwood H. Wiggins, Jr. Stanford: Stanford University Press, 2012.

Campe, Rüdiger. "An Outline for a Critical History of Fürsprache: Synegoria and Advocacy." *Deutsche Vierteljahrsschrift für Literaturwissenschaft und Geistesgeschichte* 82, no. 3 (2008): 355–381. https://doi.org/10.1007/BF03374707.

Canovan, Margaret. *Hannah Arendt: A Reinterpretation of her Political Thought*. Cambridge: Cambridge University Press, 1992.

Caputo, John D. "Demythologizing Heidegger: 'Alētheia' and the History of Being." *The Review of Metaphysics* 41, no. 3 (1988), 519–546. https://doi.org/10.2307/20128629.

Carnap, Rudolf. "Überwindung der Metaphysik durch logische Analyse der Sprache." *Erkenntnis* 2 (1931): 219–241. https://doi.org/10.2307/20011640.

Carroll, Noel. "On the Narrative Connection." In *New Perspectives on Narrative Perspective*, edited by Willie van Peer and Seymour Chatman, 21–42. Albany: SUNY Press, 2001.

Caygill, Howard. *Walter Benjamin: The Colour of Experience*. London: Routledge, 1998.

Caygill, Howard. "Walter Benjamin's Concept of Cultural History." In *The Cambridge Companion to Walter Benjamin*, edited by David S. Ferris, 73–96. Cambridge: Cambridge University Press, 2004.

Charles, Sébastien. "Le possible comme critique du Spinozisme: Leibniz et la fiction." *Science et Esprit: Revue de Philosophie et de Théologie* 67, no. 1 (2015): 17–33.

Conrad, Elfriede. *Kants Logikvorlesungen als neuer Schlüssel zur Architektonik der Kritik der reinen Vernunft: Die Ausarbeitung der Gliederungsentwürfe in den Logikvorlesungen als Auseinandersetzung mit der Tradition*. Stuttgart: frommann-holzboog, 1994.

Culler, Jonathan. "Phenomenology and Structuralism." *Human Context* 5 (1973): 35–42.

Currie, Mark. *The Unexpected: Narrative Temporality and the Philosophy of Surprise*. Edinburgh: Edinburgh University Press, 2013.

Dahlstrom, Daniel O. *The Heidegger Dictionary*. London: Bloomsbury, 2013.

De Briey, Laurent. "Le formalisme pratique: De la morale à l'éthique." *Philosophiques* 32, no. 2 (2005): 319–342. https://doi.org/10.7202/011870ar.

De Buzon, Frédéric. "Littérature et Fiction: Leibniz et Malebranche." *Dix-septième Siècle* 255, no. 2 (2012): 241–256. https://doi.org/10.3917/dss.122.0241.

Deiters, Franz-Josef. "From Collective Creativity to Authorial Primacy: Gottsched's Reformation of the German Theatre from a Mediological Point of View." In *Collective Creativity: Collaborative Work in the Sciences, Literature and the Arts*, edited by Gerhard Fischer and Florian Vassen, 75–86. Amsterdam: Rodopi, 2011.

Deligiorgi, Katerina. "The Role of the 'Plan of Nature' in Kant's Account of History from a Philosophical Perspective." *British Journal for the History of Philosophy* 14, no. 3 (2006): 451–468. https://doi.org/10.1080/09608780600794899.

Deng, Xiaomang. "Heidegger's Distortion of Dialectics in *Hegel's Concept of Experience*." *Frontiers of Philosophy in China* 4, no. 2 (2009): 294–307. https://doi.org/10.2307/40343925.

Derrida, Jacques. *The Truth in Painting*. Chicago: University of Chicago Press, 1987.

Domanska, Ewa. "Frank Ankersmit: From Narrative to Experience." *Rethinking History* 13, no. 2 (2009): 175–195. https://doi.org/10.1080/13642520902833809.

Downing, Lisa. "Occasionalism and Strict Mechanism: Malebranche, Berkeley, Fontenelle." In *Early Modern Philosophy: Mind, Matter, and Metaphysics*, edited by Christia Mercer and Eileen O'Neill, 206–230. Oxford: Oxford University Press, 2005.

Dray, William. *Philosophical Analysis and History*. New York: Greenwood Press, 1966.

Durst, Margarete. "Birth and Natality in Hannah Arendt." In *Analecta Husserliana: The Yearbook of Phenomenological Research 79*, edited by Anna-Teresa Tymieniecka, 777–797. Dordrecht: Kluwer, 2004.

Emad, Parvis. "Translating Heidegger's *Beiträge zur Philosophie* as an Hermeneutic Responsibility." *Studia Phaenomenologica: Romanian Journal of Phenomenology* 6 (2006): 347–368. https://www.ceeol.com/search/viewpdf?id=177977.

Fackenheim, Emil. *Metaphysics and Historicity*. Milwaukee: Marquette University Press, 1961.

Fagniez, Guillaume. "Hermeneutik im Übergang von Dilthey zu Heidegger." *Studia Phaenomenologica: Romanian Journal of Phenomenology* 13 (2013): 429–447. https://www.ceeol.com/search/viewpdf?id=256252.

Faulkner, William. "Banquet Speech." *Nobelprize.org*. Nobel Media AB 2014. Web. 5 Jun 2017. http://www.nobelprize.org/nobel_prizes/literature/laureates/1949/faulkner-speech.html.

Feldman, Karen S. "Not Dialectical Enough: On Benjamin, Adorno and Autonomous Critique." *Philosophy and Rhetoric* 44, no. 4 (2011): 336–362.

Feldman, Karen S. "On the Performative Difficulty of Being and Time." *Philosophy Today* 44, no. 4 (2000): 366–379. https://doi.org/10.5840/philtoday200044423.

Feldman, Karen S. "On Vitality, Figurality, and Orality in Hannah Arendt." In *Thinking Allegory Otherwise*, edited by Brenda Machowsky, 237–248. Stanford: Stanford University Press, 2013.

Fenves, Peter D. *Arresting Language: From Leibniz to Benjamin*. Stanford: Stanford University Press, 2001.

Fenves, Peter D. *The Messianic Reduction: Walter Benjamin and the Shape of Time*. Stanford: Stanford University Press, 2011.

Ferrari, G.W.F. "Aristotle's Literary Aesthetics." *Phronesis* 64, no. 3 (1999): 181–198. https://doi.org/10.1163/15685289960500024.

Ferris, David. *Theory and the Evasion of History*. Baltimore: The Johns Hopkins University Press, 1993.

Fiorato, Pierfrancesco. "'Zeitlos und dennoch nicht ohne historischen Belang': Über die idealen Zusammenhänge der Geschichte bei dem jungen Benjamin und Hermann Cohen." *MLN* 127, no. 3 (2012): 611–624. https://doi.org/10.1353/mln.2012.0096.

Fleming, Paul. "On the Edge of Non-Contingency: Anecdotes and the Lifeworld." *Telos* 158 (2012): 21–35, https://doi.org/10.3817/0312158021.

Fleming, Paul. "'Kannitverstan': The Contingent Understanding of Anecdotes." *Oxford German Studies* 40, no. 1 (2011): 72–81. https://doi.org/10.1179/007871911x568061.

Fleming, Paul. "The Perfect Story: Anecdote and Exemplarity in Linnaeus and Blumenberg." *Thesis Eleven* 104, no. 1 (2011): 72–86. https://doi.org/10.1177/0725513610394736.

Forman, David. "*Appetimus Sub Ratione Boni*: Kant's Practical Principles between Crusius and Leibniz." In *Kant und die Philosophie in Weltbürgerlicher Absicht*, edited by Stefano Bacin et al., 323–334. Berlin: De Gruyter, 2015.

Foucault, Michel. "What is Enlightenment?" In *The Foucault Reader*, edited by Paul Rabinow, 32–50. New York: Pantheon Books, 1984.

Fraser, Nancy. "Hannah Arendt in the Twenty-First Century." *Contemporary Political Theory* 3, no. 3 (2004): 253–261. https://doi.org/10.1057/palgrave.cpt.9300183.

Freier, Hans. *Kritische Poetik: Legitimation und Kritik der Poesie in Gottscheds Dichtkunst*. Stuttgart: Metzler 1973.

Freytag, Wiebke. "Die Fabel als Allegorie: Zur poetologischen Begriffssprache der Fabeltheorie von der Spätantike bis ins 18. Jahrhundert (II)." *Mittellateinisches Jahrbuch: Internationale Zeitschrift für Mediavistik* 20 (1985): 66–102.

Fulda, Daniel. "Literary Criticism and Historical Science: The Textuality of History in the Age of Goethe – and Beyond." In *The Discovery of Historicity in German Idealism and Historism*, edited by Peter Koslowski, 112–133. Berlin: De Gruyter, 2005.

Fulda, Daniel. *Wissenschaft aus Kunst: Die Entstehung der modernen deutschen Geschichtsschreibung 1760–1860*. Berlin: De Gruyter, 1996.

Fynsk, Christopher. *Heidegger: Thought and Historicity*. Ithaca: Cornell University Press, 1993.

Gadamer, Hans-Georg. "Martin Heidegger's One Path." In *Reading Heidegger from the Start: Essays in his Earliest Thought*, edited by Theodore Kisiel and John van Buren, 19–35. Albany: SUNY Press, 1994.

Gadamer, Hans-Georg. *Truth and Method*. Translated by Joel Weinsheimer and Donald G. Marshall. New York: Continuum, 2012.

Gallagher, Catherine. "The Formalism of Military History." *Representations* 104 (2008): 23–33. https://doi.org/10.1525/rep.2008.104.1.23.

Gallagher, Catherine. "The Rise of Fictionality." In *The Novel*, edited by Franco Moretti, 336–363. Princeton: Princeton University Press, 2006.

Gallagher, Catherine. "War, Counterfactual History, and Alternate-History Novels." *Field Day Review* 3 (2007): 52–65. http://www.jstor.org/stable/30078840.

Gebhard, Walter Payter. "Die Grundlagen der deutschen Fabeldichtung des 16. und 18. Jahrhunderts." In *Fabelforschung*, edited by Peter Hasubek, 268–336. Darmstadt: Wissenschaftliche Buchgesellschaft, 1983.

Geiger, Ido. "What is the Use of the Universal Law Formula of the Categorical Imperative?" *British Journal for the History of Philosophy* 18, no. 2 (2010): 271–295. https://doi.org/10.1080/09608781003643568.

Geyl, Pieter. *From Ranke to Toynbee: Five Lectures on Historians and Historiographical Problems*. Northampton, Mass.: Smith College, 1952.

Gilgen, Peter. *Lektüren der Erinnerung: Lessing, Hegel, Kant*. Munich: Fink, 2012.

Golden, Leon. *Aristotle's Poetics: A Translation and Commentary for Students of Literature*. Tallahassee: Florida State University Press, 1981.
Golden, Leon. "The Purgation Theory of Catharsis." *The Journal of Aesthetics and Art Criticism* 31, no. 4 (1973): 473–479. https://doi.org/10.2307/429320.
Gombocz, István. "'Es ist keine Wissenschaft von seinem Bezirke ganz ausgeschlossen': Johann Christoph Gottsched und das Ideal des aufklärerischen *Poeta Doctus*." *Daphnis: Zeitschrift fur Mittlere Deutsche Literatur* 18, no. 3 (1989): 541–561.
Goodman, Katherine R. "Gottsched's Literary Reforms: The Beginning of Modern German Literature." In *German Literature of the Eighteenth Century: The Enlightenment and Sensibility*, edited by Barbara Becker-Cantarino, 55–78. Rochester: Camden House, 2005.
Gordon, Peter Eli. "Science, Finitude, and Infinity: Neo-Kantianism and the Birth of Existentialism." *Jewish Social Studies* 6, no. 1 (1999): 30–53. https://doi.org/10.2979/JSS.1999.6.1.30.
Gottsched, Johann Christoph. *Ausführliche Redekunst: Nach Anleitung der alten Griechen und Römer, wie auch der neuern Ausländer, in zweenen Theilen verfasset und itzo mit den Zeugnissen der Alten und Exempeln der grössten deutschen Redner erläutert*. Leipzig: Breitkopf, 1759.
Gottsched, Johann Christoph. *Versuch einer critischen Dichtkunst*. Vol. 6, bk. 1 of *Ausgewählte Werke*, edited by Joachim Birke and Brigitte Birke. Berlin: De Gruyter, 1973.
Gräfe, Ulf. "Die rationalistische Kontrolle der Metapher in der kritischen Poetik Gottscheds." In *Kommunikative Metaphorik: Die Funktion des literarischen Bildes in der deutschen Literatur von ihren Anfangen bis zur Gegenwart*, edited by Holger A. Pausch, 81–95. Bonn: Bouvier, 1976.
Green, Leon. *Rationale of Proximate Cause*. Kansas City, MO: Vernon Law Book Company, 1927.
Grondin, Jean. "L'aletheia entre Platon et Heidegger." *Revue de Metaphysique et de Morale* 87, (1982): 551–556. https://doi.org/10.2307/40902425.
Guignon, Charles. "History and Historicity." In *A Companion to Phenomenology and Existentialism*, edited by Hubert L. Dreyfus and Mark A. Wrathall, 545–558. Oxford: Blackwell, 2008.
Guyer, Paul. "Ends of Reason and Ends of Nature: The Place of Teleology in Kant's Ethics." In *Kant's System of Nature and Freedom: Selected Essays*, 169–197. Oxford: Oxford University Press, 2005.
Guyer, Paul. *Kant on Freedom, Law, and Happiness*. Cambridge: Cambridge University Press, 2000.
Habermas, Jürgen. "Martin Heidegger: Werk und Weltanschauung." In *Texte und Kontexte*, 49–83. Frankfurt am Main: Suhrkamp, 1991.
Habermas, Jürgen. "Work and Weltanschauung: The Heidegger Controversy from a German Perspective." In *The New Conservatism: Cultural Criticism and the Historians' Debate*. Edited and translated by Shierry Weber Nicholsen, 140–172. Cambridge: MIT Press, 1989.
Hanssen, Beatrice. *Walter Benjamin and the Arcades Project*. London: Bloomsbury, 2006.
Hanssen, Beatrice. *Walter Benjamin's Other History: Of Stones, Animals, Human Beings, and Angels*. Berkeley: University of California Press, 1998.
Hardison, O.B., Jr. "Commentary on Aristotle's *Poetics*." In *Aristotle's Poetics: A Translation and Commentary for Students of Literature*, 53–296. Tallahassee: Florida State University Press, 1981.

Hassinger, Erich. "'Historische' Erkenntnis in der frühen Neuzeit." *Historische Zeitschrift* 226, no. 41 (1978): 89–101. http://www.jstor.org/stable/27620493.

Härter, Andreas. *Digressionen: Studien zum Verhältnis von Ordnung und Abweichung in Rhetorik und Poetik: Quintilian, Opitz, Gottsched, Friedrich Schlegel*. Paderborn: Fink, 2000.

Härter, Andreas, "Die Rhetorik der 'verblümten Redensarten' in Gottscheds *Versuch einer Critischen Dichtkunst*." *Colloquium Helveticum: Cahiers Suisses de Littérature Comparée/ Schweizer Hefte für Allgemeine und Vergleichende Literaturwissenschaft/Quaderni Svizzeri di Letteratura Generale e Comparata* 30 (1999): 25–44.

Haverkamp, Anselm. "The Scandal of Metaphorology." *Telos* 158 (2012): 37–58. https://doi.org/10.3817/0312158037.

Hegel, Georg Wilhelm Friedrich. *Lectures on the Philosophy of World History, Introduction: Reason in History*. Translated by H.B. Nisbet. Cambridge: Cambridge University Press, 1975.

Hegel, Georg Wilhelm Friedrich. *Vorlesungen über die Philosophie der Geschichte*. Werke, vol. 12, edited by Eva Moldenhauer and Karl Markus Michel. Frankfurt am Main: Suhrkamp, 1986.

Heidegger, Martin. "Anmerkungen zu Karl Jaspers 'Psychologie der Weltanschauungen.'" In *Wegmarken*, 1–44.

Heidegger, Martin. *Basic Concepts of Aristotelian Philosophy*. Translated by Robert D. Metcalf and Mark B. Tanzer. Bloomington: Indiana University Press, 2009.

Heidegger, Martin. *Basic Problems of Phenomenology*. Translated by Albert Hofstadter. Bloomington: Indiana University Press, 1988.

Heidegger, Martin. *Basic Questions of Philosophy: Selected "Problems" of "Logic"*. Translated by Richard Rojcewicz and André Schuwer. Albany: SUNY Press, 1994.

Heidegger, Martin. *Being and Time*. Translated by Joan Stambaugh. Albany: SUNY Press, 1997.

Heidegger, Martin. "Comments on Karl Jaspers' *Psychology of Worldviews*." Translated by John van Buren. In *Pathmarks*, 1–38.

Heidegger, Martin. "On the Essence of Ground." Translated by William McNeill. In *Pathmarks*, 97–135.

Heidegger, Martin. "On the Essence of Truth (1930)." Translated by John H. Sallis. In *Pathmarks*, 136–154.

Heidegger, Martin. *Grundbegriffe der aristotelischen Philosophie. Gesamtausgabe*, vol. 18, edited by Mark Michalski. Frankfurt am Main: Klostermann, 2002.

Heidegger, Martin. *Grundfragen der Philosophie: Ausgewählte 'Probleme' der 'Logik.' Gesamtausgabe*, vol. 45, edited by Friedrich-Wilhelm von Herrmann. Frankfurt am Main: Klostermann, 1984.

Heidegger, Martin. *Die Grundprobleme der Phänomenologie. Gesamtausgabe*, vol. 24, edited by Friedrich-Wilhelm von Herrmann. Frankfurt am Main: Klostermann, 1975.

Heidegger, Martin. "The Origin of the Work of Art." In *Off the Beaten Track*, edited and translated by Julian Young and Kenneth Haynes, 1–56. Cambridge: Cambridge University Press, 2002.

Heidegger, Martin. *Ontologie: Hermeneutik der Faktizität. Gesamtausgabe*, vol. 63, edited by Käte Bröcker-Oltmanns. Frankfurt am Main: Klostermann, 1988.

Heidegger, Martin. *Ontology: The Hermeneutics of Facticity*. Translated by John van Buren. Bloomington: Indiana University Press, 1999.

Heidegger, Martin. *Pathmarks*. Edited by Will McNeill. Cambridge: Cambridge University Press, 1998.

Heidegger, Martin. *The Principle of Reason*. Translated by Reginald Lilly. Bloomington: Indiana University Press, 1996.
Heidegger, Martin. *Sein und Zeit. Gesamtausgabe*, vol. 2, edited by Friedrich-Wilhelm von Herrmann. Frankfurt am Main: Klostermann, 1977.
Heidegger, Martin. "Der Ursprung des Kunstwerks." In *Holzwege. Gesamtausgabe*, vol. 5, edited by Friedrich-Wilhelm von Herrmann, 7–68. Frankfurt am Main: Klostermann, 1963.
Heidegger, Martin. "Vom Wesen der Wahrheit." In *Wegmarken*, 177–202.
Heidegger, Martin. "What is Metaphysics?" In *Pathmarks*, 82–96.
Heidegger, Martin. *Wegmarken. Gesamtausgabe*, vol. 9, edited by Friedrich-Wilhelm von Herrmann. Frankfurt am Main: Klostermann, 1967.
Heidegger, Martin. "Wilhelm Diltheys Forschungsarbeit und der gegenwärtige Kampf um eine historische Weltanschauung." In *Reden und andere Zeugnisse eines Lebensweges 1910–1976. Gesamtausgabe*, vol. 16, edited by Hermann Heidegger, 49–51. Frankfurt am Main: Klostermann, 2000.
Heidegger, Martin. "Wilhelm Dilthey's Research and the Current Struggle for a Historical Worldview (1925)." In *Becoming Heidegger: On the Trail of His Early Occasional Writings, 1910–1927*, edited by Theodore Kisiel and Thomas Sheehan, 238–274. Evanston: Northwestern University Press, 2007.
Hemming, Laurence Paul. "Speaking Out of Turn: Martin Heidegger and die Kehre." *International Journal of Philosophical Studies* 6, no. 3 (1998): 393–423, https://doi.org/10.1080/096725598342046.
Henrich, Dieter. *The Course of Remembrance and Other Essays on Hölderlin*. Edited by Eckart Förster. Translated by Taylor Carman. Stanford: Stanford University Press, 1997.
Herder, Johann Gottfried. *Auch eine Philosophie der Geschichte zur Bildung der Menschheit: Beitrag zu vielen Beiträgen des Jahrhunderts*. Frankfurt am Main: Suhrkamp, 1967.
Herman, Barbara. "Embracing Kant's Formalism." *Kantian Review* 16, no. 1 (2011): 49–66. https://doi.org/10.1017/S1369415410000075.
Herman, Barbara. "A Habitat for Humanity." In *Kant's Idea for a Universal History with a Cosmopolitan Aim: A Critical Guide*. Edited by James Schmidt and Amélie Rorty, 150–170. Cambridge: Cambridge University Press, 2009.
Hesse, Carla. "A Fugitive Book." *Representations* 104 (2008): 37–49. https://doi.org/10.1525/rep.2008.104.1.37.
Hinske, Norbert. *Zwischen Aufklärung und Vernunftkritik: Studien zum Kantschen Logikcorpus*. Stuttgart: frommann-holzboog, 1998.
Hoffmann, Stefan-Ludwig. "Koselleck, Arendt, and the Anthropology of Historical Experience." *History and Theory* 49, no. 2 (2010): 212–236.
Hoffmann, Stefan-Ludwig, and Sean Franzel. Introduction to *Sediments of Time: On Possible Histories*, edited and translated by Sean Franzel and Stefan-Ludwig Hoffmann, ix–xxxi. Stanford: Stanford University Press, 2018.
Holborn, Hajo. *History and the Humanities*. Garden City, New York: Doubleday, 1972.
Holenstein, Elmar. "Jakobson and Husserl: A Contribution to the Genealogy of Structuralism." *Human Context* 7 (1975): 61–83.
Hume, David. *A Treatise of Human Nature*. Oxford: Oxford University Press, 2000.
Humboldt, Wilhelm von. "Die Aufgabe des Geschichtsschreibers." In *Wilhelm von Humboldts Gesammelte Schriften*, vol. 4, edited by the Königlich Preußische Akademie der Wissenschaften, 35–56. Berlin: Behr, 1903.
Humboldt, Wilhelm von. "On the Historian's Task." *History and Theory* 6, no. 1 (1967): 57–71.

Husain, Martha. *Ontology and the Art of Tragedy: An Approach to Aristotle's Poetics*. Albany: SUNY Press, 2012.
Ingarden, Roman. *The Literary Work of Art*. Evanston: Northwestern University Press, 1973.
Jackson, S.W. "Catharsis and Abreaction in the History of Psychological Healing." *Psychiatric Clinics of North America* 17, no. 3 (1994): 471–491.
Jakobson, Roman. "Modern Russian Poetry: Velimir Khlebnikov." In *Major Soviet Works: Essays in Criticism*, edited by E.J. Brown, 58–82. Oxford: Oxford University Press, 1973.
Jakobson, Roman. "Zur Struktur des Phonems." In *Selected Writings I: Phonological Studies*, 280–310. Berlin: De Gruyter, 2012.
Jameson, Fredric. "The Aesthetics of Singularity." *New Left Review* 92 (March–April 2015): 101–132. https://newleftreview.org/II/92/fredric-jameson-the-aesthetics-of-singularity.
Jameson, Fredric. *The Antinomies of Realism*. London: Verso, 2015.
Janko, Richard. *Aristotle on Comedy: Towards a Reconstruction of Poetics II*. Berkeley: University of California Press, 1984.
Jaszczolt, Kasia. *Representing Time: An Essay on Temporality as Modality*. Oxford: Oxford University Press, 2009.
Jauss, Hans Robert. "Literary History as a Challenge to Literary Theory." In *Toward an Aesthetic of Reception*, translated by Timothy Bahti, 3–45. Minneapolis: University of Minnesota Press, 1982.
Jordheim, Helge. "Against Periodization: Koselleck's Theory of Multiple Temporalities." *History and Theory* 51 (2012): 151–171.
Kaelin, Eugene F. "Language as a Medium for Art." *Journal of Aesthetics and Art Criticism* 40 (1981): 121–130. https://doi.org/10.2307/430404.
Kang, Taran. "Origin and Essence: The Problem of History in Hannah Arendt." *Journal of the History of Ideas* 74, no. 1 (2013): 139–160. https://doi.org/10.2307/23354926.
Kant, Immanuel. "Conjectures on the Beginnings of Human History." In *Political Writings*, 221–234.
Kant, Immanuel. "Contest of the Faculties." Excerpted in *Political Writings*, 177–190.
Kant, Immanuel. *Critique of the Power of Judgment*. Edited by Paul Guyer. Translated by Paul Guyer and Eric Matthews. Cambridge: Cambridge University Press, 2000.
Kant, Immanuel. *Critique of Pure Reason*. Edited and translated by Paul Guyer and Allen W. Wood. Cambridge: Cambridge University Press, 1998.
Kant, Immanuel. *Gesammelte Schriften*. Volumes 1–22 edited by the Preußische Akademie der Wissenschaften, volume 23 by the Deutsche Akademie der Wissenschaften zu Berlin, and volumes 24 and beyond by the Akademie der Wissenschaften zu Göttingen. Berlin: Reimer/De Gruyter, 1900–. Hereafter abbreviated AA, followed by volume and page numbers.
Kant, Immanuel. "Idea for a Universal History with a Cosmopolitan Purpose." In *Political Writings*, 41–53.
Kant, Immanuel. "Idee zu einer allgemeinen Geschichte in weltbürgerlicher Absicht." AA 8:15–32.
Kant, Immanuel. *Kritik der reinen Vernunft*. AA 3.
Kant, Immanuel. *Kritik der Urteilskraft*. AA 5:167–485.
Kant, Immanuel. *Lectures on Logic*. Edited and translated by J. Michael Young. Cambridge: Cambridge University Press, 1992.
Kant, Immanuel. "Logik Blomberg." AA 24, no. 1, 16–301.
Kant, Immanuel. *Logik Nachlaß*. AA 16.

Kant, Immanuel. "Mutmaßlicher Anfang der Menschengeschichte." AA 8:107–124.
Kant, Immanuel. *Notes and Fragments*. Edited by Paul Guyer. Translated by Curtis Bowman, Paul Guyer, and Frederick Rauscher. Cambridge: Cambridge University Press, 2005.
Kant, Immanuel. *Political Writings*. Edited by Hans Reiss. Cambridge: Cambridge University Press, 1991.
Kant, Immauel. "Reviews of Herder's *Ideas on the Philosophy of the History of Mankind*." In *Political Writings*, 201–220.
Kant, Immanuel. "Rezensionen von J.G. Herders Ideen zur Philosophie der Geschichte der Menschheit." AA 8:43–66.
Kant, Immanuel. "Der Streit der Fakultäten." AA 7:1–116.
Kateb, George. *Hannah Arendt: Politics, Conscience, Evil*. Totowa, NJ: Rowman and Allanheld, 1983.
Katz, Jacob. *Exclusiveness and Tolerance: Studies in Jewish-Gentile Relations in Medieval and Modern Times*. London: Oxford University Press, 1961.
Kayser, Wolfgang Peter. "Die Grundlagen der deutschen Fabeldichtung des 16. und 18. Jahrhunderts." In *Fabelforschung*, edited by Peter Hasubek, 79–96. Darmstadt: Wissenschaftliche Buchgesellschaft, 1983.
Kelley, Donald R. Review of *Cognitio Historica: Die Geschichte als Namengerberin der frühneuzeitlichen Empirie*, by Arno Seifert, and other works. *The Journal of Modern History* 54, no. 2 (1982): 320–326. http://www.jstor.org/stable/1906161.
Kermode, Frank. "Secrets and Narrative Sequence." In *On Narrative*, edited by W.J.T. Mitchell, 79–97. Chicago: University of Chicago Press, 1981.
Kermode, Frank. *The Sense of An Ending: Studies in the Theory of Fiction*. Oxford: Oxford University Press, 2000.
Kittsteiner, Heinz Dieter. *Out of Control: Über die Unverfügbarkeit des historischen Prozesses*. Berlin: Philo, 2004.
Kittsteiner, Heinz Dieter. "Walter Benjamin's Historicism." *New German Critique* 39 (Fall 1986): 179–215. https://doi.org/10.2307/488125.
Klaus, Peter. "Der gute Herrscher: Literarische Beispiele bei Gottsched und in der Romantik." In *Geschichtlichkeit und Aktualität: Studien zur deutschen Literatur seit der Romantik*, edited by Klaus-Detlef Müller et al., 97–112. Tübingen: Niemeyer, 1988.
Kleingeld, Pauline. "Kant on Historiography and the Use of Regulative Ideas." *Studies in History and Philosophy of Science* 39, no. 4 (2008): 523–528. https://doi.org/10.1016/j.shpsa.2008.09.006.
Koschorke, Albrecht. "Codes und Narrative: Überlegungen zur Poetik der funktionalen Differenzierung." In *Grenzen der Germanistik: Rephilologisierung oder Erweiterung?*, edited by Walter Erhart, 174–185. Stuttgart: Metzler, 2004. https://doi.org/10.1007/978-3-476-05570-5_11.
Koschorke, Albrecht. "In Praise of the Undefined: Toward a General Theory of Narrativity." Lecture at the Wissenschaftskolleg zu Berlin, 2011.
Koschorke, Albrecht. *Wahrheit und Erfindung: Grundzüge einer Allgemeinen Erzähltheorie*. Frankfurt am Main: Fischer, 2012.
Koselleck, Reinhart. "Das achtzehnte Jahrhundert als Beginn der Neuzeit." In *Epochenschwelle und Epochenbewusstsein*, edited by Reinhart Herzog and Reinhart Koselleck, 269–282. Munich: Fink, 1987.
Koselleck, Reinhart. "Chance as Motivational Trace in Historical Writing." In *Futures Past*, 116–129.

Koselleck, Reinhart. "On the Disposability of History." In *Futures Past*, 198–212.
Koselleck, Reinhart. "The Eighteenth Century as the Beginning of Modernity." In *The Practice of Conceptual History*, 154–169.
Koselleck, Reinhart. "'Erfahrungsraum' und 'Erwartungshorizont': Zwei historische Kategorien." In *Vergangene Zukunft: Zur Semantik geschichtlicher Zeiten*, 349–374.
Koselleck, Reinhart. *Futures Past: On the Semantics of Historical Time*. Translated by Keith Tribe. Cambridge: MIT Press, 1985.
Koselleck, Reinhart. "Geschichte, Historie." In *Geschichtliche Grundbegriffe: Historisches Lexikon zur politisch-sozialen Sprache in Deutschland*, edited by Otto Brunner, Werner Conze, and Reinhart Koselleck, 593–717. Stuttgart: Klett, 1975.
Koselleck, Reinhart. "Historia Magistra Vitae: The Dissolution of the Topos into the Perspective of a Modernized Historical Process." In *Futures Past*, 21–38.
Koselleck, Reinhart. "Historik und Hermeneutik." *Sitzungsberichte der Heidelberger Akademie der Wissenschaften* 1 (1987): 9–28.
Koselleck, Reinhart. "History, Histories, and Formal Structures of Time." In *Futures Past*, 92–104.
Koselleck, Reinhart. "*Neuzeit:*Remarks on the Semantics of Modern Concepts of Movement." In *Futures Past*, 231–266.
Koselleck, Reinhart. "Perspective and Temporality: A Contribution to the Historiographical Exposure of the Historical World," in *Futures Past*, 130–158.
Koselleck, Reinhart. *The Practice of Conceptual History: Timing History, Spacing Concepts*. Translated by Todd Presner, Kerstin Behnke and Jobst Welge. Stanford: Stanford University Press, 2002.
Koselleck, Reinhart. *Sediments of Time: On Possible Histories*. Edited and translated by Sean Franzel and Stefan-Ludwig Hoffmann. Stanford: Stanford University Press, 2018.
Koselleck, Reinhart. "'Space of Experience' and 'Horizon of Expectation': Two Historical Categories." In *Futures Past*, 267–288.
Koselleck, Reinhart. "Terror and Dream: Methodological Remarks on the Experience of Time during the Third Reich." In *Futures Past*, 213–230.
Koselleck, Reinhart. *Vergangene Zukunft: Zur Semantik geschichtlicher Zeiten*. Frankfurt am Main: Suhrkamp Verlag, 1995.
Koselleck, Reinhart. *Zeitschichten: Studien zur Historik*. Frankfurt am Main: Suhrkamp Verlag, 2000.
Kosik, Karel. "Historism and Historicism." *New German Critique* 10 (1977): 65–75. https://doi.org/10.2307/487672.
Krell, David Farrell. "On the Manifold Meaning Of Aletheia: Brentano, Aristotle, Heidegger." *Research in Phenomenology* 5 (1975): 77–94. https://doi.org/10.2307/24654276.
Krieger, Leonard. "Elements of Early Historicism: Experience, Theory and History in Ranke." *History and Theory* 14, no. 4 (1975): 1–14. https://doi.org/10.2307/2504662.
Kuehn, Manfred. *Reason as a Species Characteristic*. Cambridge: Cambridge University Press, 2009. https://doi.org/10.1017/CBO9780511581434.005.
LaCapra, Dominick. "Kant, Benjamin, Pensky and the Historical Sublime." *The Philosophical Forum* 41, nos. 1–2 (2010): 175–179. https://doi.org/10.1111/j.1467-9191.2009.00357.x.
Lakoff, George. "Global Warming *Systemically* Caused Hurricane Sandy," on *Berkeley Blog*, November 5, 2012. http://blogs.berkeley.edu/2012/11/05/global-warming-systemically-caused-hurricane-sandy/.
Laqueur, Thomas. "Form in Ashes." *Representations* 104 (2008): 50–72. https://doi.org/10.1525/rep.2008.104.1.50.

Lausberg, Heinrich. *Handbook of Literary Rhetoric*. Edited by David E. Orton and R. Dean Anderson. Translated by Matthew T. Bliss, Annemiek Jansen, and David E. Orton. Leiden: Brill, 1998.

Lear, Jonathan. "Katharsis." In *Essays on Aristotle's Poetics*, edited by Amélie Rorty, 315–340. Princeton: Princeton University Press, 1992.

Leibniz, G.W. *Theodicy: Essays on the Goodness of God, the Freedom of Man, and the Origin of Evil*. Translated by Austin Farrer. La Salle, IL: Open Court, 1985.

Lepenies, Wolf. *Das Ende der Naturgeschichte: Wandel kultureller Selbstverständlichkeiten in der Wissenschaft des 18. und 19. Jahrhunderts*. Munich: Hanser, 1976.

Lessing, Gotthold Ephraim. "Briefe, die neueste Literatur betreffend." *Werke*, vol. 5, edited by Herbert G. Göpfert et al., 30–329. Munich: Hanser, 1978.

Lessing, Gotthold Ephraim. *Hamburg Dramaturgy*. Translated by Helen Zimmern. New York: Dover, 1962.

Lessing, Gotthold Ephraim. *Hamburgische Dramaturgie. Werke*, vol. 4, edited by Herbert G. Göpfert et al., 229–707. Munich: Hanser, 1979.

Levinson, Marjorie. "What is the New Formalism?" *PMLA* 122, no. 2 (2007): 558–569. https://doi.org/10.1632/pmla.2007.122.2.558.

Lorenz, Sarah Ruth. "Shifting Forms of Mimesis in Johann Christoph Gottsched's *Dichtkunst*." *German Quarterly* 87, no. 1 (Winter 2014): 86–107. https://doi.org/10.1111/gequ.10199.

Lucas, F.L. *Tragedy in Relation to Aristotle's* Poetics. London: Hogarth Press, 1927.

Lukacher, Ned. *Primal Scenes: Literature, Philosophy, Psychoanalysis*. Ithaca: Cornell University Press, 1986.

Luther, Martin. "The Freedom of a Christian." In *Martin Luther's Basic Theological Writings*, edited by Timothy F. Lull, 585–629. Minneapolis: Fortress Press, 1989.

Lyotard, Jean-François. *Heidegger and "the jews."* Translated by Andreas Michel and Mark Roberts. Minneapolis: University of Minnesota Press, 1990.

Mackie, J.L. *The Cement of the Universe: A Study of Causation*. Oxford: Clarendon Press, 1974.

Makkreel, Rudolf. *Imagination and Interpretation in Kant: The Hermeneutical Import of the Critique of Judgment*. Chicago: University of Chicago Press, 1990.

Makkreel, Rudolf A. "The Overcoming of Linear Time in Kant, Dilthey, and Heidegger." In *Dilthey and Phenomenology*, edited by Rudolf A. Makkreel, 141–158. Lanham: University Press of America, 1987.

Makkreel, Rudolf A. "Purposiveness in History: Its Status after Kant, Hegel, Dilthey and Habermas." *Philosophy & Social Criticism* 18 (1992): 221–234. https://doi.org/10.1177/019145379201800301.

Malinowski-Charles, Sylviane. "De la possibilité des fictions littéraires chez Spinoza." *Teoria: Rivista di Filosofia* 32, no. 2 (2012): 247–265. https://www.academia.edu/13811564/Spinoza_et_les_fictions_vraies.

Marcuse, Herbert. *Hegel's Ontology and the Theory of Historicity*. Translated by Seyla Benhabib. Cambridge: MIT Press, 1987.

Martyn, David. "Figures of the Mean: Freedom, Progress, and the Law of Statistical Averages in Kleist's 'Allerneuester Erziehungsplan.'" *The Germanic Review: Literature, Culture, Theory* 85, no. 1 (2010): 44–62. https://doi.org/10.1080/00168890903446674.

McCarthy, Michael H. *The Political Humanism of Hannah Arendt*. Lanham MD: Lexington Books, 2014.

McCumber, John. *The Company of Words: Hegel, Language, and Systematic Philosophy*. Evanston: Northwestern University Press, 1993.

McMahon, Jenny. "The Classical Trinity and Kant's Aesthetic Formalism." *Critical Horizons* 11, no. 3 (2010): 419–441. https://doi.org/10.1558/crit.v11i3.419.

Meier, Georg Friedrich. *Excerpt from the Doctrine of Reason*. Translated by Aaron Bunch. London: Bloomsbury, 2016.

Mensch, Jennifer. *Kant's Organicism: Epigenesis and the Development of Critical Philosophy*. Chicago: University of Chicago Press, 2013.

Merritt, Melissa McBay. "Reflection, Enlightenment, and the Significance of Spontaneity in Kant." *British Journal for the History of Philosophy* 17, no. 5 (2009): 981–1010. https://doi.org/10.1080/09608780903339178.

Miller, J. Hillis. "Narrative." In *Critical Terms for Literary Study*, edited by Frank Lentricchia and Thomas McLaughlin, 66–79. Chicago: University of Chicago Press, 1990.

Mitchell, Andrew J. "The Coming of History: Heidegger and Nietzsche against the Present." *Continental Philosophy Review* 46, no. 3 (2014): 395–411. https://doi.org/10.1007/s11007-013-9269-6.

Mitchell, P.M. *Johann Christoph Gottsched (1700–1766): Harbinger of German Classicism*. Columbia, S.C.: Camden House, 1995.

Mormann, Thomas and Mikhail Katz, "Infinitesimals as an Issue of Neo-Kantian Philosophy of Science." *HOPOS: The Journal of the International Society for the History of Philosophy of Science* 3, no. 2 (2013): 236–280. https://doi.org/10.1086/671348.

Morson, Gary Saul. *Narrative and Freedom: The Shadows of Time*. New Haven: Yale University Press, 1994.

Möller, Uwe. *Rhetorische Überlieferung und Dichtungstheorie im frühen 18. Jahrhundert*. Paderborn: Fink, 1983.

Murphy, Sinéad. *Effective History: On Critical Practice under Historical Conditions*. Evanston: Northwestern University Press, 2010.

Nadler, Steven M. "Occasionalism and General Will in Malebranche." *Journal of the History of Philosophy* 31, no. 1 (1993): 31–47.

Naragon, Steve. "Kant in the Classroom: Materials to Aid the Study of Kant's Lectures." http://www.manchester.edu/kant/notes/notesIntro.htm.

Nehamas, Alexander. "Pity and Fear in the *Rhetoric* and *Poetics*." In *Essays on Aristotle's Poetics*, edited by Amélie Rorty, 291–315. Princeton: Princeton University Press, 1990.

Nelson, Eric S. "Heidegger and Dilthey: Language, History, and Hermeneutics." In *Horizons of Authenticity in Phenomenology, Existentialism, and Moral Psychology*, edited by Hans Pedersen and Megan Altman, 109–128. Dordrecht: Springer, 2015.

Ng, Julia. "Walter Benjamin's and Gershom Scholem's Reading Group around Hermann Cohen's *Kants Theorie der Erfahrung* in 1918: An Introduction." *MLN* 127, no. 3 (2012): 433–461. https://doi.org/10.1353/mln.2012.0085.

Nietzsche, Friedrich. *Unfashionable Observations*. Translated by Richard T. Gray. Stanford: Stanford University Press, 1995.

Nietzsche, Friedrich. *Unzeitgemäße Betrachtungen. Werke in drei Bänden*, vol. 1, edited by Karl Schlechta. Munich: Carl Hanser Verlag, 1954.

Nirenberg, David. *Anti-Judaism: The Western Tradition*. New York: Norton, 2013.

Norkus, Zenonas. "Why Hannah Arendt's Ideas on Totalitarianism Are Heterodox." *Topos: Journal for Philosophical and Cultural Studies* 2, no. 19 (2008): 114–136.

North, Joseph. *Literary Criticism: A Concise Political History*. Cambridge: Harvard University Press, 2017.

Nussbaum, Martha C. *Cultivating Humanity: A Classical Defense of Reform in Liberal Education*. Cambridge: Harvard University Press, 1998.
Nussbaum, Martha C. *The Fragility of Goodness: Luck and Ethics in Greek Tragedy and Philosophy*. Cambridge: Cambridge University Press, 2013.
Nuzzo, Angelica. *Kant and the Unity of Reason*. Lafayette: Purdue University Press, 2005.
Olay, Csaba. "Die Überlieferung der Gegenwart und die Gegenwart der Überlieferung: Heidegger und Gadamer über Tradition." *International Yearbook for Hermeneutics* 12 (2013): 196–219.
Otter, Samuel. "An Aesthetics in All Things." *Representations* 104 (2008): 116–125. https://doi.org/10.1525/rep.2008.104.1.116.
Pensky, Max. "Contributions Toward a Theory of Storms: Historical Knowing and Historical Progress in Kant and Benjamin." *The Philosophical Forum* 41, nos. 1–2 (2010): 149–174. https://doi.org/10.1111/j.1467-9191.2009.00356.x.
Perler, Dominik and Ulrich Rudolph. *Occasionalismus: Theorien der Kausalität im arabisch-islamischen und im europäischen Denken*. Göttingen: Vandenhoeck & Ruprecht, 2000.
Peters, Brigitte. "Der 17. Literaturbrief und seine Folgen." *Zeitschrift für Germanistik* 10, no. 1 (1989): 70–75. http://www.jstor.org/stable/23974739.
Petrusevski, M.D. "La définition de la tragédie chez Aristote et la catharsis" *L'Annuaire de la Faculté de philosophie de l'Université de Skopje* 1 (1948): 3–17.
Petrusevski, M.D. "Pathematon Katharsin ou bien Pragmaton Systasin?" *Ziva Antika/Antiquité Vivante* 4, no. 2 (1954): 209–250.
Pippin, Robert B. "Kant on the Spontaneity of Mind." In *Hegelian Variations: Idealism as Modernism*, 29–55. Cambridge: Cambridge University Press, 1997.
Plug, Jan. "'As-if' History – Kant, Then and Now" in *Borders of a Lip: Romanticism, Language, History, Politics*, 17–44. Albany: SUNY Press, 2012.
Popper, Karl. *The Open Society and its Enemies*. London: Routledge, 2011.
Popper, Karl. *The Poverty of Historicism*. New York: Harper, 1961.
Pozzo, Riccardo. *Kant und das Problem einer Einleitung in die Logik: Ein Beitrag zur Rekonstruktion der historischen Hintergründe von Kants Logik-Kolleg*. Frankfurt am Main: Lang, 1989.
Pöggeler, Otto. "'Historicity' in Heidegger's Late Work." *Southwestern Journal of Philosophy* 4 (1973): 53–73. https://doi.org/10.2307/43154948.
Pugliese, Orlando. *Vermittlung und Kehre: Grundzüge des Geschichtsdenkens bei Martin Heidegger*. Freiburg: Alber, 1965.
Ranke, Leopold von. "Critique of Guicciardini." In *Secret of World History*, 73–98.
Ranke, Leopold von. "Einleitung zu einer Vorlesung über Universalhistorie." Edited by Eberhard Kessel. *Historische Zeitschrift* 176 (1954): 304–307.
Ranke, Leopold von. "Epochs of Modern History." In *Secret of World History*, 156–169.
Ranke, Leopold von. *Geschichten der romanischen und germanischen Völker von 1494 bis 1515*. In *Sämtliche Werke*, vols. 33–34.
Ranke, Leopold von. "The Historian's Ideal." In *Secret of World History*, 259.
Ranke, Leopold von. "The Historian's Task." In *Secret of World History*, 258.
Ranke, Leopold von. "History and Philosophy." In *Secret of World History*, 101–104.
Ranke, Leopold von. *Über die Epochen der neueren Geschichte: Vorträge dem Könige Maximillian II. von Bayern im Herbst 1854 zu Berchtesgaden gehalten*, edited by Theodor Schieder and Helmut Berding. Munich: Oldenbourg, 1971.
Ranke, Leopold von. *Zur Kritik neuer Geschichtsschreiber*, vols. 33–34, *Sämtliche Werke*.

Ranke, Leopold von. "The Pitfalls of a Philosophy of History." In *Theory and Practice of History*, 17–19.
Ranke, Leopold von. "Power and Spiritual Force in History." In *Secret of World History*, 257.
Ranke, Leopold von. "Preface to the First Edition of *Histories of the Latin and Germanic Peoples*." In *Theory and Practice of History*, 85–87.
Ranke, Leopold von. "On Progress in History (From the first lecture to King Maximilian II of Bavaria 'On the Epochs of Modern History, 1854')." In *Theory and Practice of History*, 20–23.
Ranke, Leopold von. "On the Relation of and Distinction between History and Politics." In *Theory and Practice of History*, 75–82.
Ranke, Leopold von. "On the Relations of History and Philosophy." In *Theory and Practice of History*, 5–7.
Ranke, Leopold von. *Sämmtliche Werke*. Edited by Alfred Wilhelm Dove and Theodor Wiedemann. 54 vols. Leipzig, Duncker & Humblot, 1873–1890
Ranke, Leopold von. *The Secret of World History: Selected Writings on the Art and Science of History*. Edited and translated by Roger Wines. New York: Fordham University Press, 1981.
Ranke, Leopold von. *The Theory and Practice of History*. Edited by Georg G. Iggers. Translated by Wilma A. Iggers. New York: Routledge, 2011.
Ranke, Leopold von. "Über die Verwandtschaft und den Unterschied der Historie und der Politik." In *Sämtliche Werke*, vol. 24, 280–293.
Reinhardt, Mark. "What's New in Arendt?" *Political Theory: An International Journal of Political Philosophy* 31, no. 3 (2003): 443–460. https://doi.org/10.2307/3595683.
Ricci, Gabriel. *Time Consciousness: The Philosophical Uses of History*. New Brunswick: Transaction, 2010.
Richter, Gerhard. *Afterness: Figures of Following in Modern Thought and Aesthetics*. New York: Columbia University Press, 2011.
Ricoeur, Paul. *Time and Narrative*. Chicago: The University of Chicago Press, 1988.
Rorty, Richard. *Contingency, Irony, and Solidarity*. New York: Cambridge University Press, 1989.
Rudnick, Hans H. "Roman Ingarden's Literary Theory." *Analecta Husserliana* 4 (1976): 105–119. https://doi.org/10.1007/978-94-010-1443-4_3.
Russell, Bertrand. *Human Knowledge*. New York: Simon and Schuster, 1948.
Rüsen, Jörn. "Rhetoric and Aesthetics of History: Leopold von Ranke." *History and Theory* 29, no. 2 (1990): 190–204.
Sala, Giovanni B. "Der Formalismus in der Ethik Kants: Überlegungen zu einer alten Kontroverse." *Freiburger Zeitschrift für Philosophie und Theologie* 52, nos. 1–2 (2005): 191–215.
Sallis, John H. "Grounders of the Abyss." In *Companion to Heidegger's* Contributions to Philosophy, edited by Charles E. Scott et al., 181–197. Bloomington: Indiana University Press, 2001.
Sayers, Dorothy L. "Aristotle on Detective Fiction." *English* 1, no. 1 (1936): 23–35. https://doi.org/10.1093/english/1.1.23.
Scharff, Robert C. "Heidegger's 'Appropriation' of Dilthey before *Being and Time*." *Journal of the History of Philosophy* 35, no. 1 (1997): 105–128. https://doi.org/10.1353/hph.1997.0021.
Schatzki, Theodore R. "Living Out of the Past: Dilthey and Heidegger on Life and History." *Inquiry* 46, no. 3 (2003): 301–323. https://doi.org/10.1080/00201740310002389.
Schäfer, Gerhard. *'Wohlklingende Schrift' und 'rührende Bilder': Soziologische Studien zur Ästhetik Gottscheds und der Schweizer*. Frankfurt am Main: Lang, 1987.

Scholl, Rosemary. "Die Rhetorik der Vernunft: Gottsched und die Rhetorik im frühen 18. Jahrhundert." *Jahrbuch für Internationale Germanistik* 2, no. 3 (1976): 217–221.
Schwebel, Paula L. "Intensive Infinity: Walter Benjamin's Reception of Leibniz and its Sources." *MLN* 127, no. 3 (2012): 589–610. https://doi.org/10.1353/mln.2012.0093.
Scott, Gregory. "Purging the *Poetics*." *Oxford Studies in Ancient Philosophy* 25 (Winter 2003): 233–263.
Sedgwick, Sally. "Hegel on the Empty Formalism of Kant's Categorical Imperative." In *A Companion to Hegel*, edited by Stephen Houlgate, 265–280. Oxford: Blackwell, 2011.
Seifert, Arno. *Cognitio Historica: Die Geschichte als Namengeberin der frühneuzeitlichen Empirie*. Berlin: Duncker & Humblot, 1976.
Selinger, William. "The Politics of Arendtian Historiography: European Federation and *The Origins of Totalitarianism*." *Modern Intellectual History* 13, no. 2 (2016): 417–446. https://doi.org/10.1017/S1479244314000560.
Setton, Dirk. "Absolute Spontaneity of Choice: The Other Side of Kant's Theory of Freedom." *Symposium* 17, no. 1 (2013): 75–99. https://doi.org/10.5840/symposium20131715.
Sgarbi, Marco. *Kant on Spontaneity*. London: Continuum, 2012.
Sheehan, Thomas. "Geschichtlichkeit/Ereignis/Kehre." *Existentia: An International Journal of Philosophy* 11, nos. 3–4 (2001): 241–51.
Shklar, Judith N. "Rethinking the Past." *Social Research* 44, no. 1 (1977): 80–90. https://doi.org/10.2307/40970272.
Silber, John R. "Procedural Formalism in Kant's Ethics." *The Review of Metaphysics* 28, no. 2 (1974): 197–236. https://doi.org/10.2307/20126622.
"Silva Rhetoricae/The Forest of Rhetoric" (website). Rhetoric.byu.edu.
Silverman, Hugh. *Inscriptions: Between Phenomenology and Structuralism*. New York: Routledge, 1987.
Smith, John H. "The Infinitesimal as Theological Principle: Representing the Paradoxes of God and Nothing in Cohen, Rosenzweig, Scholem, and Barth." *MLN* 127, no. 3 (2012): 562–588. https://doi.org/10.1353/mln.2012.0090.
Smith, John H. "Kant, Calculus, Consciousness, and the Mathematical Infinite in Us." *Goethe Yearbook* 23 (2016): 95–121. https://doi.org/10.1353/gyr.2016.0001.
Starn, Randolph. "Historicizing Representations: A Formal Exercise." *Representations* 104 (2008): 137–143. https://doi.org/10.1525/rep.2008.104.1.137.
Staudenmaier, Peter. "Hannah Arendt's Analysis of Antisemitism in *The Origins of Totalitarianism*: A Critical Appraisal." *Patterns of Prejudice* 46, no. 2 (2012): 154–179. https://doi.org/10.1080/0031322X.2012.672224.
Stauffer, Hermann. *Erfindung und Kritik: Rhetorik im Zeichen der Frühaufklärung bei Gottsched und seinen Zeitgenossen*. Frankfurt am Main: Lang, 1997.
Stawarska, Beate. *Saussure's Philosophy of Language as Phenomenology: Undoing the Doctrine of the Course in General Linguistics*. Oxford: Oxford University Press, 2015.
Straßberger, Andres. *Johann Christoph Gottsched und die "philosophische" Predigt: Studien zur aufklärerischen Transformation der protestantischen Homiletik im Spannungsfeld von Theologie, Philosophie, Rhetorik und Politik*. Tübingen: Mohr Siebeck, 2010.
Striedter, Jurij. *Literary Structure, Evolution and Value*. Translated by Matthew Gurewitch. Cambridge: Harvard University Press, 1989.
Sweet, Paul R. "Wilhelm von Humboldt (1767–1835): His Legacy to the Historian." *The Centennial Review* 15, no. 1 (1971): 23–37. https://doi.org/10.2307/23737762.

Sznaider, Natan. "Hannah Arendt and the Sociology of Antisemitism." *Österreichische Zeitschrift für Politikwissenschaft* 39, no. 4 (2010): 421–434. https://doi.org/10.15203/ozp.586.vol39iss4.

Szondi, Peter. *An Essay on the Tragic*. Translated by Paul Fleming. Stanford: Stanford University Press, 2002.

Terada, Rei. "Thinking for Oneself: Realism and Defiance in Arendt." *ELH* 71, no. 4 (2004): 839–865. https://doi.org/10.1353/elh.2004.0051.

Tonner, Philip. "Epoch: Heidegger and the Happening of History." *Minerva: An Open Access Journal of Philosophy* 19 (2015): 132–150. http://www.minerva.mic.ul.ie/Vol19/History.pdf.

Trüper, Henning. "Löwith, Löwith's Heidegger, and the Unity of History." *History and Theory* 53 (2014): 45–68. https://doi.org/10.1111/hith.10694.

Tynjanov, Jurij. "On Literary Evolution." In *Readings in Russian Poetics: Formalist and Structuralist Views*, edited by Ladislav Matejka and Krystyna Pomorska, 66–78. Champaign: Dalkey Archive Press, 2002.

Uhlig, Claus. "Literature as Textual Palingenesis: On Some Principles of Literary History." *New Literary History* 16, no. 3 (1985): 481–513. https://doi.org/10.2307/468837.

Vaihinger, Hans. *Die Philosophie des Als Ob: System der theoretischen, praktischen und religiösen Fiktionen der Menschheit auf Grund eines idealistischen Positivismus*. Leipzig: Meiner, 1922.

Veloso, Claudio William. "Aristotle's *Poetics* without Katharsis, Fear or Pity." *Oxford Studies in Ancient Philosophy* 33 (Winter 2007): 255–284.

Vierhaus, Rudolf. "Historiography between Science and Art." In *Leopold von Ranke and the Shaping of the Historical Discipline*, edited by Georg G. Iggers and James M. Powell, 61–69. Syracuse: Syracuse University Press, 1990.

Villa, Dana. "Political Violence and Terror: Arendtian Reflections." *Ethics and Global Politics* 1, no. 3 (2008): 97–113. https://doi.org/10.3402/egp.v1i3.1861.

Vowinckel, Annette. *Arendt*. Leipzig: Reclam, 2006.

Wallwitz, Georg. "Kant über Fatalismus und Spontaneität." *Allgemeine Zeitschrift für Philosophie* 28, no. 3 (2003): 207–228.

Walsh, David. Introduction to *Anamnesis: On the Theory of History and Politics*, by Eric Voegelin, 1–27. Columbia: University of Missouri Press, 2002.

Wand, Bernard. "Religious Concepts and Moral Theory: Luther and Kant." *Journal of the History of Philosophy* 9, no. 3 (1971): 329–348. https://doi.org/10.1353/hph.2008.1268.

Weber, Samuel. *Benjamin's –abilities*. Cambridge: Harvard University Press, 2010.

Weber, Samuel. "Psychoanalysis and Theatricality." *Parallax* 6, no. 3 (2000). doi:10.1080/135346400422448.

Weber, Samuel. *Theatricality as Medium*. New York: Fordham University Press, 2004.

Wellbery, David. "Aesthetic Media: The Structure of Aesthetic Theory before Kant." In *Regimes of Description: In the Archive of the Eighteenth Century*, edited by John Bender and Michael Marrinan, 199–215. Stanford: Stanford University Press, 2005.

Wellmon, Chad. *Becoming Human: Romantic Anthropology and the Embodiment of Freedom*. University Park, PA: Pennsylvania State University Press, 2010.

White, Hayden. "Appendix on Narration, Narrative, Narrativization." In *The Practical Past*, 93–96. Evanston: Northwestern University Press, 2014.

White, Hayden. Foreword to *The Practice of Conceptual History: Timing, Spacing, Concepts*, by Reinhart Koselleck, ix–xiv. Stanford: Stanford University Press, 2002.

White, Hayden. "Historical Discourse and Literary Writing." In *Tropes for the Past: Hayden White and the History/Literature Debate*, edited by Kuisma Korhonen, 25–35. Amsterdam: Rodopi, 2006.

White, Hayden. "Introduction: Historical Fiction, Fictional History, and Historical Reality." *Rethinking History* 9, nos. 2–3 (2005): 147–157. https://doi.org/10.1080/13642520500149061.

White, Hayden. "The Irrational and the Problem of Historical Knowledge in the Enlightenment." In *Studies in Eighteenth-Century Culture: Proceedings of the American Society for Eighteenth-Century Studies*, edited by Harold E. Pagliaro, 303–321. Cleveland: Case Western Reserve University Press, 1972.

White, Hayden. *Metahistory: The Historical Imagination in Nineteenth-Century Europe*. Baltimore: The Johns Hopkins University Press, 1973.

White, Hayden. *Tropics of Discourse: Essays in Cultural Criticism*. Baltimore: The Johns Hopkins University Press, 1978.

White, Hayden. "The Value of Narrativity in the Representation of Reality." In *The Content of the Form: Narrative Discourse and Historical Representation*, 1–25. Baltimore: The Johns Hopkins University Press, 1987.

Williams, Eric. *The Mirror and the Word: Modernism, Literary Theory, and Georg Trakl*. Lincoln: University of Nebraska Press, 1993.

Williams, Seán M. "Kant's Novel Interpretation of History." *Seminar: A Journal of Germanic Studies* 49, no. 2 (2013): 171–190. https://doi.org/10.3138/sem.49.2.171.

Wimmer, Mario. *Archivkörper: Eine Geschichte historischer Einbildungskraft*. Konstanz: Konstanz University Press, 2012.

Wrathall, Mark A. *Heidegger and Unconcealment: Truth, Language, and History*. Cambridge: Cambridge University Press, 2011.

Wren, Thomas E. "Heidegger's Philosophy of History." *Journal of the British Society for Phenomenology* 3 (1972): 111–125. https://doi.org/10.1080/00071773.1972.11006246.

Wurzer, Jörg. "Der Freiheitsbegriff Martin Luthers und Immanuel Kants im Vergleich." *Luther: Zeitschrift der Luther-Gesellschaft* 71, no. 1 (2000): 21–35.

Young-Bruehl, Elisabeth. *Hannah Arendt: For Love of the World*. New Haven: Yale University Press, 1982.

Zambrana, Rocío. "Hegel's Hyperbolic Formalism." *Hegel Bulletin* 31, no. 1 (2010): 107–131. https://doi.org/10.1017/S0263523200001099.

Zammito, John H. "Koselleck's Philosophy of Historical Time(s) and the Practice of History." *History and Theory* 43 (2004): 124–135. https://doi.org/10.2307/3590747.

Zammito, John H. "A Text of Two Titles: Kant's 'A Renewed Attempt to Answer the Question: "Is the Human Race Continually Improving?"' *Studies in History and Philosophy of Science Part A* 39, no. 4 (2008): 535–545. https://doi.org/10.1016/j.shpsa.2008.09.004.

Zangwill, Nick. "Feasible Aesthetic Formalism." *Nous* 33, no. 4 (1999): 610–629. https://doi.org/10.2307/2671957.

Zerilli, Linda. "Castoriadis, Arendt, and the Problem of the New." *Constellations* 9, no. 4 (2002): 540–553.

Zuckert, Rachel. "The Purposiveness of Form: A Reading of Kant's Aesthetic Formalism." *Journal of the History of Philosophy* 44, no. 4 (2006): 599–622. https://doi.org/10.1353/hph.2006.0075.

Index

adnominatio 124
Adorno, Theodor 95, 98, 99, 104, 110, 114
adversatives (e.g., but, still, yet) 30
aesthetic judgment 79, 80–81
aesthetics 33, 101, 106, 170. *See also* aesthetic judgment; beauty
– Gottsched and 6, 38, 46
– history and 11, 65n16, 98–99, 103–104, 108, 112–114, 172n6
– Humboldt and 93
– in Kant's logic lectures 71–77
– Kantian universal history and 11, 79, 82
– tragedy and 18, 23, 24, 27, 29, 30
affect/affectability 155, 156, 173. *See also pathos/pathe;* fear; pity
Alcibiades 20
allegorization 12, 120, 126, 129, 131, 134
Amthor, Christoph Heinrich 51–56
anamnesis 120, 126, 127
Anderson, John G.T. 71n21
Ankersmit, Frank 83n13, 172n6
anthimeria 124
anti-Semitism 138–142, 143
Arendt, Hannah 3, 93, 119, 135–152, 153, 154, 173
– "The Concept of History" 138, 147, 148, 149, 150, 151
– *The Human Condition* 136
– Introduction to *Between Past and Future* 149
– *On Revolution* 136, 137n8
– *Origins of Totalitarianism* 12–13, 35–152, 171
Aristotle 1, 2–3, 8, 41, 128, 156
– Gottsched and 6, 10, 38–42, 43n14, 47
– *Metaphysics* 2n6, 21
– *Poetics* 1, 9–10, 17–37, 38, 40, 41n11, 42, 43n14, 170
Armstrong, J.M. 17, 19–21, 27, 33n58, 36
arrangement of events. *See plot*
as-if
– Arendt and 136, 147, 149, 151, 164
– Aristotle's *Poetics* and 34
– Gottsched and 56, 57

– Kant's universal history and 84, 136, 151
– narrativization and 168n48
– Vaihinger on Kant's 57, 93
Atomism 90
Auerbach, Erich 149
Augustine, Saint 144
awe 30, 31, 32

Baehr, Peter 140n13
Bahti, Timothy 79n4
Baudelaire, Charles 110
Baumgarten, Alexander 72, 75, 77, 156, 170
Bayle, Pierre 91, 92
beauty 33n57, 53, 54, 71–75, 80–82, 102. *See also* aesthetic judgment; aesthetics
Bede, the Venerable [Saint] 156
beginnings. *See* freedom; newness; spontaneity
Begriffsgeschichte 13, 64, 76, 153–154, 157–163, 169
Beiser, Frederick 38, 44, 74, 75n25, 100, 106
Benjamin, Walter 4, 46
– Arendt and 114, 148, 151
– Kant and 79–80, 94–97
– Ranke and 11, 92, 98–99, 103–114
Bennett, Benjamin 39n5
Berghahn, Klaus 42n12
Bernstein, Richard J. 129n23
Blackall, Eric 43n13, 48
Bloch, Ernst 166
Blomberg, Hermann Ulrich von 63
Bloom, Harold 11n21
Blumenberg, Hans 5, 7n17, 33n59, 48n21, 155n7
Bodmer, Johann Jakob 39n5, 44n14, 50
borrowing 40–44, 56
Brandt, Reinhard 63
Breitinger, Johann Jakob 39n5, 50
Brobjer, Thomas 114n65
Brodsky, Claudia 23–24n23
Brooks, Peter 1n4, 170, 171
Brunner, Otto 153
Burckhardt, Jakob 109

Burke, Edmund 33n57
Burke, Peter 100, 111, 114

Campe, Rüdiger 40–41, 42n12, 49, 50, 155–156, 163n35, 173
Caputo, John 128
Carnap, Rudolf 124
Cassirer, Ernst 95, 153n2
catharsis 9, 21–26, 36
causality 1–3, 8, 9, 11, 13, 66, 111. *See also* continuity; freedom; necessity; one-by-one connection of events; process
– in Arendt 140, 145, 147, 151
– in Aristotle 9, 26, 30–32
– in Gottsched 3, 48–54
– in Kant 11, 87–90, 97, 135
– in Koselleck 13, 160, 163
– in Leibniz 33
– in Ranke 110n49
Caygill, Howard 95n60, 96n63, 109n47
chance 31, 32, 33n58, 162–163, 165. *See also* contingency
chiasmus 43, 44
Christianity, Christians and 87, 102, 127, 150n33
– Arendt on Jews and 139–142
Cicero 48, 48n23, 156, 157
cognitio historica. *See* historical knowledge
Cohen, Hermann 94, 95
completeness 1n4, 44, 109, 149, 150n31, 174. *See also* beauty; continuity; unity; wholeness
– knowledge and (in Kant and Meier) 68, 71–75
– tragedy and 17, 18, 30, 32, 33n58, 170
– of the *Poetics* 35, 36
– White and 169n48, 172, 173
connexio realis 40–42, 48–51, 55–57
connexio verbalis 48, 49
Conrad, Elfriede 62n3
contingency 4–9, 11–13, 97, 155n7, 163n35, 170, 171
– Arendt and 134, 145, 146, 147, 150
– tragedy and 17, 18
continuity 3, 6–8, 10, 171
– in Arendt 134, 137, 138–142, 144, 146, 149, 151, 152

– Aristotle's *Poetics* and 33, 34, 36
– in Blumenberg 4–5
– in Heidegger 12, 119, 122, 129–132
– in Kant 6, 56, 70, 79, 88n33, 89–99
– in Koselleck 158–160, 163–166
– in Ranke 6, 105, 106, 114
Conze, Werner 153
Currie, Mark 6n16

Darius the Great 164
Davila, Arrigo Caterino 103
deconstruction 6, 101, 162, 165, 166, 167, 169
Deligiorgi, Katerina 84n16
Dilthey, Wilhelm 121, 123, 126, 127n14, 150, 153n2
dispositio 1
Dornblüth, August 48, 48n23
Droysen, Johann Gustav 11, 114n65

Ellison, Ralph 24
emotion. *See* affect; fear; *pathos/pathe*; pity
emplotment 1, 8, 17. *See also* plot
– Arendt and 146–147
– Aristotle and 9, 21, 27
– Heidegger and 3, 12, 119, 123, 126–127, 131, 132–134
– Kant and 98
– White and 105
Enlightenment 7, 140
– Gottsched and 46n19, 56
– Kant and 65, 76, 87
– tragedy and 23, 24, 27, 38
epistemology 61, 70, 98, 99, 108, 111, 160
epochs/epochality 3, 5, 6, 8, 12, 101, 171, 172
– Arendt and 13, 135–152
– Heidegger and 12, 13, 119–120, 126–134, 157
– Koselleck and 153–169
– Ranke and 99, 105–107, 108, 112, 114–115
examples/exemplarity 9, 112, 169
– Koselleck and 13, 76, 107, 153–164, 167–168

Fabel 3, 10, 39–40, 47–50, 55–57. *See also* plot
Fackenheim, Emil 88n34, 101

Faulkner, William 19
fear 9, 18, 21–26, 30, 32, 156
feeling. See aesthetic judgment; aesthetics; affect; beauty; catharsis; fear; *pathos/pathe*; pity
Fenves, Peter 97n70
Ferrari, G.W.F. 27–32, 34, 36
Ferris, David 34, 36
Fichte, Johann Gottlieb 106
figure, figurality. See adnominatio; allegorization; anthimeria; chiasmus; dispositio; hendiadys; hypotyposis; metaphor; metonymy; polyptoton; taxis
Fiorato, Pierfrancesco 95n57
Fleming, Paul 7n17, 155n7
Florus, Publius Annius 43
formalism 4, 14, 81, 82, 102, 109n45. See also New Formalism
Forman, David 88n33
Foucault, Michel 87
Franzel, Sean 158n20
Frederick the Great 168
freedom 5–9, 92, 93, 94, 95, 108n41, 171–173. See also causality; spontaneity
– Arendt and 12, 13, 93, 134, 135–152
– Heidegger and 127, 132–134
– Kant and 11, 79–83, 85, 87–90, 97
Freud, Sigmund 22, 35, 127n15
Fulda, Daniel 65

Gadamer, Hans-Georg 86, 157, 158
Gallagher, Catherine 102n17, 172n6
Genette, Gérard 171n5
Geschichtlichkeit. See historicity
Gilgen, Peter 87n29
God, gods, and goddesses 32, 33, 88, 90, 102, 139, 162
– every epoch immediate to (Ranke) 99, 105, 106
– in Gottsched 47, 49, 53, 54
– the novel and 90, 91, 92
Golden, Leon 24, 25, 26, 27, 30, 36
Gordon, Peter Eli 95n57
Gottsched, Johann Christoph 3, 8, 61, 171
– Aristotle and 2, 6, 9–10, 38–42, 47

– *Ausführliche Redekunst* 48n23
– *Erste Gründe der gesammten Weltweisheit* 43n18
– *Versuch einer critischen Dichtkunst* 38–57
Guicciardini, Francesco 11, 104–105, 108, 110n53, 112
Guyer, Paul 87n29

Habermas, Jürgen 129–131
Halliwell, Stephen 17n1, 20, 30
Hanssen, Beatrice 110n49
Hardison, O.B. 20n12, 24, 25, 26, 27, 36
Haverkamp, Anselm 4n8, 5n15
Hegel, Georg Wilhelm Friedrich 10, 12, 81, 98, 100–102, 148
– Ranke and 106, 107, 112–114
Heidegger, Martin 3, 12–13, 119–134, 171, 173
– Arendt and 13, 136–137, 150
– Koselleck and 153, 154, 157–158, 161, 169
– *Basic Questions of Philosophy* 127–128, 132
– *Being and Time* 119, 121, 123
– *Black Notebooks* 129
– "Comments on Karl Jaspers' *Psychology of Worldviews*" 121–122, 125
– "On the Essence of Ground" 124
– "On the Essence of Truth" 12, 120, 125
– "Letter on Humanism" 131
– National Socialism and 129–131
– "The Origin of the Work of Art" 12, 120, 125
– *The Principle of Reason* 128
– "What is Metaphysics?" 124
– "Wilhelm Dilthey's Research and the Current Struggle for a Historical Worldview" 121, 123, 126, 150
Heimsoeth, Heinz 94
hendiadys 23
Henrich, Dieter 133n29
Herder, Johann Gottfried von 62, 92n45, 99, 100n9, 106n34, 157, 166n43
– Ranke and 106, 107, 108
Herman, Barbara 86
hermeneutics 76, 86, 107, 153, 154, 161
Herodotus 149, 155
Hippocrates 69, 70

historical knowledge (*cognitio historica*) 10, 61–77, 78, 94, 170, 172
historicism 5, 14, 92–94
– Ranke and 3, 11–12, 98–103, 106–107, 110, 111–115
historicity (*Geschichtlichkeit*) 12, 119–134, 150, 154, 157–158, 171. *See also* Heidegger
historiography. *See* literary writing, comparison to historiography
Hoffmann, Stefan-Ludwig 158n20, 160, 161
Holborn, Hajo 106
Homer 74, 75, 148, 149
Humboldt, Wilhelm von 76, 92–94, 103, 109
Hume, David 2n6
Husain, Martha 25, 26
Husserl, Edmund 12, 97n70, 119, 121, 137
hypotyposis 155

ideology 3, 101, 105, 107, 109, 113
– Arendt and 141, 144, 147
– White and 132, 133
imagination 7, 8, 13, 24, 170
– Gottsched and 10, 43, 45
– Humboldt and 93
– Kant and 77, 78–79, 80–82
– Ranke and 113–115
– tragedy and 28, 29, 31
Ingarden, Roman 137
"interstitial" connection. *See* one-by-one connection of events
Isidore of Seville, Saint 156

Jakobson, Roman 101, 137
Jameson, Fredric 106, 107n41
Janko, Richard 30
Jaspers, Karl 121–123, 125, 126
Jauss, Hans Robert 5
Jordheim, Helge 161, 162

Kästner, Abraham Gotthelf 46
Kant, Immanuel 3, 4, 6, 7, 8, 56–57, 135, 154, 161, 170
– Blomberg logic (lectures on logic) 10, 61–77, 78, 171
– "Conjectures on the Beginning of Human History" 92
– *Contest of the Faculties* 83, 89
– *Critique of the Power of Judgment* 33n57, 62, 79, 81–82, 84, 93, 94
– *Critique of Practical Reason* 11, 88, 97, 151
– "Idea for a Universal History with a Cosmopolitan Purpose" 10–11, 78–97, 98, 100–102, 107–108, 136, 137, 146, 151, 167–168, 172
– *Perpetual Peace* 96
– "What is Enlightenment?" 87
Kateb, George 145
katharsis. *See* catharsis
Katz, Jacob 139, 140, 141
Kermode, Frank 2, 149
Kittsteiner, Heinz Dieter 79n4, 92n45, 99, 107
Kleingeld, Pauline 86n24
Koschorke, Albrecht 6n16, 65n15, 79n4
Koselleck, Reinhart 3, 12, 13, 112, 119, 153–169, 170, 171, 173
– "Chance as a Motivational Trace" 162–163, 168
– "On the Disposability of History" 165
– "The Eighteenth Century as the Beginning of Modernity" 166
– "Geschichte, Historie" 154–157
– *Geschichtliche Grundbegriffe* 153
– "Historia Magistra Vitae" 40, 64, 76, 93n48, 107, 154–159, 168
– "Historik und Hermeneutik" 158
– "History, Histories, and Formal Structures of Time" 154, 156, 164
– "Neuzeit: Remarks on the Semantics of Modern Concepts of Movement" 159
– "Perspective and Temporality" 159
– *The Practice of Conceptual History* 158
– "Terror and Dream" 165
– *Zeitschichten* 154, 166
Krieger, Leonard 100, 109n49, 113
Kuehn, Manfred 79n4, 91n41

La Popelinière, Lancelot Voisin de 114
Lear, Jonathan 19, 25n28
Lehmann, Gerhard 63

Leibniz, Gottfried Wilhelm 33, 80, 91n43, 92, 94, 167n43
- Heidegger and 128
- Kant and 79, 80, 88n33, 90, 91
- Malebranche and 90n39
- Wolff and 72

Lepenies, Wolf 71n21
Lessing, Gotthold Ephraim 9, 22, 23, 38
Levinson, Marjorie 101
literary criticism 1, 5, 9, 17, 32, 34, 137. See also literary theory
literary theory 1, 101. See also literary criticism
- Aristotle and 18, 30, 34, 35, 36, 37, 170
- Ranke and 98, 101.
literary writing, comparison to historiography 1, 3, 4, 6–8, 9
- in Aristotle 19
- in Gottsched 39, 40, 41–44, 50–51, 54, 56
- in Kant and Meier 64, 65, 76–77, 101, 172, 173
- in Kant and Meier 76
- in Humboldt 93
- in Kant's universal history 90–91
- in Ranke 99, 103–104, 114
Livius, Titus (Livy) 43
Lorenz, Sarah Ruth 38n1
Luther, Martin 87, 88, 92
Luxemburg, Rosa 136
Lyotard, Jean-François 119, 127n15

Maharal (Judah Loew ben Bezalel) 139
Malebranche, Nicolas 90
Marcuse, Herbert 101
Martyn, David 89n35
Marx, Karl 149
McCumber, John 102
Meier, Georg F. 10, 61–77, 172. See also Kant, Blomberg logic
Meinecke, Friedrich 106, 113
Melanchthon, Philip 156
memory 39, 45, 61, 62, 65–70, 76, 148, 150
Mendelssohn, Moses 22
Mensch, Jennifer 87n29
metaphor 4, 93, 105n31, 148, 150, 159
metaphysics 4. See also Aristotle, *Metaphysics*

- Heidegger's history of 12, 119–120, 126–134, 136–137, 157
- Leibniz and 33, 90, 91
metonymy 45
mimesis 7, 21, 28–30, 33–36, 38, 103
Mitchell, Andrew J. 126n13
Mitys 31, 33
Montesquieu 100n9
Morson, Gary Saul 8n18
muthos (μῦθος). See plot

narrativity 10, 35, 65n15, 110, 119
- in Arendt 146, 148, 152
- in Kant's *cognitio historica* 64n14, 76–77
- White and 62n6, 132–134, 157n16, 168n48, 172
National Socialism 100
- Arendt on 140n13, 146
- Heidegger and 129–131
- Koselleck and 158, 158n20
natural process. See process
nature, human 28, 76, 84, 107
nature, plan of 79, 82–90, 94, 97, 100
necessity/necessary connection 2, 5, 6–7
- in Arendt 134, 145
- in Aristotle's *Poetics* 17–19, 25, 28–29, 33n58, 40, 41n11, 170
- in Gottsched 43, 51, 55
- in Humboldt 93, 94
Nehamas, Alexander 26
neo-Kantianism 80, 94–96, 97n70
New Formalism 1, 5, 101
newness 12, 134, 135, 137–140, 142–147, 154. See also freedom; spontaneity
Ng, Julia 95n57
Nietsche, Friedrich 98, 114, 114n65, 148, 150
Nirenberg, David 141, 142
North, Joseph 101n15
novel, the 32, 49, 107n41
- Blumenberg and 4, 5, 48n21, 171
- Kant and 78–79, 90–92, 172
Nussbaum, Martha 24, 26n34
Nuzzo, Angelica 63n8

Odysseus. *See* Homer
one-by-one ("interstitial") connection of events 1–3, 5–7, 11, 12, 14, 171, 172, 174. *See also* causality; contingency; continuity; necessity; plot
– in Arendt 137
– in Aristotle 17, 33
– in Gottsched 47
– in Kant and Meier 68
– Kant's universal history and 84, 87, 89
– in Koselleck 163

Pater Bossu (René Le Bossu) 47
pathos/pathe 25, 56, 173. *See also* affect; fear; pity
Pensky, Max 86, 109n48
perfection. *See* beauty; completeness
Petrusevski, M.D. 25, 26
phenomenology 3–8, 12, 13, 172
– Arendt and 135–138, 142–144, 146–147, 151–152
– Heidegger and 119–121, 125, 126, 128–129, 131, 132–134
– Koselleck and 13, 153–154, 158–167. *See also* Hegel
Philoctetes 24
philosophy of history 2n6, 5, 8, 62, 65n16, 100, 102, 108
– Kant, Meier and 76–77
– Kant's universal history and 79n4, 80, 96–97
– Ranke and 98–99
pity 9, 18, 21–26, 30, 32. *See also pathos/pathe*; fear
plan of nature. *See* nature, plan of
Plato 127, 128, 164
plot 1–3, 5–6, 8–14, 170–174. *See also* emplotment; *Fabel*; one-by-one connection of events
– in Arendt 146–147, 152
– in Aristotle 17–21, 24–31, 32n54, 33, 35, 36, 47, 144
– in Gottsched 47, 49
– in Heidegger 119–120, 125, 131, 132–134
poetics 40, 56, 64, 76, 93n48, 99. *See also* Aristotle, *Poetics*
Polybius 156, 164

polyptoton 124
Popper, Karl 93, 100, 102
Positivism 92, 94, 98, 103, 104, 112–114
possibility 4, 13, 91n43, 96, 134, 156, 172–173
– Arendt and 134, 136, 147, 150, 151
– Aristotle and 4, 5, 6, 8
– Gottsched and 10, 38–50, 55–56
– Kant and 79
– Heidegger and 13, 126, 128, 129, 133–134
– Koselleck and 154, 158–162, 167, 169, 171
pre-Socratics 119, 127–128, 132–133, 148
process 2n6, 80, 170
– in Arendt 93, 135, 147–151
– in Aristotle 29, 33n58
– in Gottsched 39, 56
– in Heidegger 121–122, 124
– in Humboldt 94n54
– in Kant and Meier 67, 68, 75
progress 4, 96n61, 146, 164, 170
– in Arendt 151
– in Kant 3, 11, 79, 82, 83, 84–89, 91–92, 97, 109n48, 136
– in Koselleck 167–168
– in Ranke 98–99, 100, 105–108, 112

Ranke, Leopold von 3, 6, 8, 10, 11, 94, 98–118, 119, 171
– "Critique of Guicciardini" 11, 104–105, 108, 109, 110n53, 112
– "On the Epochs of Modern History" 98, 106
– "The Historian's Ideal" 111
– "The Historian's Task" 103, 108
– "History and Philosophy" 29, 105
– "The Pitfalls of a Philosophy of History" 112, 113, 114
– "Power and Spiritual Force in History" 108
– "Preface to the First Edition of Histories of the Latin and Germanic Peoples" 99, 105, 107, 110
– "On the Relation of and Distinction between History and Politics" 103, 105, 109, 111
– "On the Relations of History and Philosophy" 106
Ranke, Otto 111

rationality 10–11, 13, 74, 75n25, 128
– in Arendt 136, 151
– emotion and 22–24
– in Gottsched 38–39, 44, 45, 46
– in Meier and Kant 61–62, 63, 65–68, 71–73, 76
– in Kant 56–57, 79n4, 82–84, 85, 87n29, 88–89, 91, 95–97, 100
reason. *See* rationality
Ricci, Gabriel 127
Richter, Gerhard 84n15
Rorty, Richard 24
Rüsen, Jörn 172n6
Rufus, Quintus Curtius 43
Russell, Bertrand 2n6
Russian formalism. *See* formalism

Sartre, Jean-Paul 107n41
Sayers, Dorothy 27
Scharfsinn/Scharfsinnigkeit
– in Gottsched 10, 39, 44–47, 50, 54, 55, 57
– in Ranke 104–105, 114
Schiller, Friedrich 79n4
Schoen, Ernst 95
Schwebel, Paula 94
Seifert, Arno 64n14, 68n18
Shaftesbury, Earl of 42n12
Shklovsky, Viktor 4
Simmel, Georg 98, 99, 104
Sleidanus, Johannes 114
Smith, John H. 94, 95n57
Sophocles 24
spontaneity 7, 8, 171
– Arendt and 13, 134, 135, 138, 144–152
– Heidegger and 134
– Kant and 11, 79, 87–89, 92, 97. *See also* freedom
Stark, Werner 63
Stauffer, Hermann 46n19
Straßberger, Andres 43n13
sublime, the 32n57, 73, 79, 82, 83, 87, 94
Sweet, Paul R. 94
synegoria 156–157, 169n48
Szondi, Peter 18

Tacitus 41
taxis 1

teleology 56
– Hegel and 100, 101, 107
– Kantian universal history and 79, 83–85, 88
– of nature 85, 94
– Ranke and 107, 109, 112, 114
temporality 1, 106, 108n41, 168n47, 170–171, 174
– in Gottsched 39, 45
– in Heidegger 119–125, 128
– in Kant and Meier 10, 61, 64, 64n14, 66–68, 70–71, 74–77
– in Kant's universal history 84, 89
– in Koselleck 13, 154, 159–160, 163–167, 169
Terada, Rei 142
Thucydides 43, 149, 156
time. *See* temporality
totalitarianism. *See* Arendt, *Origins of Totalitarianism*
tragedy 1, 9, 17–37, 43n14, 53, 144, 172
Treitschke, Heinrich von 114n65
Tynjanov, Jurij 4

unity 1–11, 63, 76, 132, 170–174. *See also* completeness; continuity; wholeness
– aesthetic 82n10
– Kantian universal history and 83–97, 146
– of the *Poetics* 32–37
– Ranke and 105, 107, 110
– of tragedy 17–18, 29, 32–33
universal history
– Arendt and 136, 147
– Hegel and 100
– Kant and 3, 10, 11, 62, 78–97, 151, 167, 172
– Koselleck and 157, 159, 166, 167, 168, 170
– Ranke's rejection of 98–99, 105n29, 106–113
universals, in tragedy 18–21, 27–28, 33, 36–37, 41n11

Vaihinger, Hans 56, 93, 151n34
Veloso, Claudio 25, 26
Vico, Giambattista 41n8
Vierhaus, Rudolf 109n45
Voegelin, Eric 128n13, 143, 144

Weber, Samuel 35, 36, 46
Wellbery, David 76
Wellmon, Chad 91n42
White, Hayden 1, 40–41, 65n16, 168n48, 172–173
– Heidegger and 132–133
– Humboldt and 92
– Kant and 62n6, 146n24
– Koselleck and 157n16, 160n28, 168n47
– Ranke and 103, 105n31, 108
wholeness 1, 2, 7, 101, 168, 170–174. *See also* completeness; sublime; unity
– Gottsched and 47, 48
– of the *Poetics* 33–37
– tragedy and 9, 17, 32, 33
– Kant, history and 78–80, 82–86, 89–94, 97
Williams, Seán 91
Wimmer, Mario 100n7
Wolff, Christian
– Gottsched and 40, 49, 50
– Kant and 88n32
– Meier and 61, 64, 66, 72, 75, 172

Xenophon 43

Zammito, John 89n37, 166n40, 167n43
Zusammenhang 47–49, 50–56, 66–67, 71, 130

www.ingramcontent.com/pod-product-compliance
Lightning Source LLC
Chambersburg PA
CBHW021730220426
43662CB00008B/786